Load Serving Entity

by
Sandeep

TABLE OF CONTENTS

DECLARATION i

CERTIFICATE iii

ACKNOWLEDGEMENTS v

ABSTRACT vii

TABLE OF CONTENTS xi

LIST OF FIGURES xvii

LIST OF TABLES xix

LIST OF SYMBOLS xxi

LIST OF ABBREVIATIONS xxv

Chapter - 1 Introduction 3

1.1 Background 3

1.2 Motivation for the Proposed Work 5

1.3 Contribution of the Proposed Work 8

1.4 Scope and Limitation 10

1.5 Thesis Organization 10

Chapter - 2 Literature Review 15

2.1 Evolving Industry Structure 15

2.2 Load Serving Entity 17

2.3 **Load Serving Entity's Operation in Electricity Markets** 17

    2.3.1 Supply Side Operation 18

    2.3.2 Demand Side Operation 20

2.4 Market Uncertainties 21

    2.4.1 Correlated Price and Demand 22

    2.4.2 Uncertainty Handling Approaches 23

        2.4.2.1 Deterministic Models 24

        2.4.2.2 Probabilistic Techniques 24

        2.4.2.3 Possibility Theory 24

        2.4.2.4 Hybrid Possibilistic-Probabilistic Techniques 24

        2.4.2.5 Stochastic Programming 25

        2.4.2.6 Information Gap Decision Theory 25

        2.4.2.7 Robust Optimization 26

        2.4.2.8 Interval Analysis 27

2.5	Risk Management Approaches	27
2.6	Consumer Behavior Modeling Approaches	28
2.6.1	Econometric Approaches	29
2.6.2	Fuzzy and Game based Approaches	29
2.6.3	Load Profile based Approaches	30
2.6.4	Price Elasticity	30
2.6.5	Bi-level Programing	31
2.7	Demand Response and its Application	31
2.7.1	Demand Response Program	32
2.7.1.1	Incentive Based Demand Response Program	32
2.7.1.2	Price Based Demand Response Program	32
2.7.2	Demand Response Contracts	33
2.7.3	Application of Demand Response	33
2.8	Impact of Retail Competition on Consumer Switching	33
2.9	Retail Pricing Schemes	34
2.9.1	Fixed Pricing	35
2.9.2	Dynamic Pricing	35
2.9.2.1	Time of Use Prices	35
2.9.2.2	Peak Prices	36
2.9.2.3	Real Time Pricing	36
2.9.3	Risk Reward Mapping of Retail Pricing	37
2.10	Impact of Renewable Energy Integration	37
2.11	Research Challenges	39

Chapter - 3 LSE's Risk Based Profit Maximization **43**

3.1	Introduction	43
3.2	Problem Description	44
3.3	Mathematical Model	45
3.3.1	Procurement Cost	45
3.3.1.1	Bilateral Contracts	46
3.3.1.2	Self-generation	46
3.3.1.3	Spot Market	48
3.3.2	Energy Balance Constraint	48
3.3.3	Revenue	48
3.3.4	Sale Price Modeling	48
3.3.5	Price Uncertainty Modeling	49

3.3.6   Objective Function 50

3.4   Case Study 51

    3.4.1   Data 51

    3.4.2   Statistical Calculations 53

    3.4.3   Simulations 54

    3.4.4   Results 54

      3.4.4.1   Case-A: Proposed Model without Renewable Generation 55

      3.4.4.2   Case-B: Proposed model with Renewable Generation 61

3.5   Conclusions 66

Chapter - 4 LSE's Risk Based Profit Maximization in Bi-level Framework 71

4.1   Introduction 71

4.2   Problem Description 72

4.3   Mathematical Model 73

    4.3.1   Upper Level: Load Serving Entity's Problem 73

      4.3.1.1   Procurement Cost 73

      4.3.1.2   Energy Balance Constraint 74

      4.3.1.3   Revenue 74

      4.3.1.4   Sale Price Modeling 75

      4.3.1.5   Price Uncertainty Modeling 75

      4.3.1.6   Load Serving Entity's Objective Function 75

    4.3.2   Lower Level: Consumers' Problem 76

      4.3.2.1   Consumers' Objective Function 76

    4.3.3   Equivalent Single-Level Bi-Level Model 77

      4.3.3.1   KKT Optimality Condition for Lower Level Problem 77

      4.3.3.2   Equivalent Linear Expression 78

    4.3.4   Overall Objective Function 79

4.4   Case Study 79

    4.4.1   Data 79

    4.4.2   Statistical Calculations 80

    4.4.3   Simulations 81

    4.4.4   Results 81

      4.4.4.1   Dynamic Sale Prices and Consumer Demand 81

      4.4.4.2   Energy Procurement 83

      4.4.4.3   Impact of Risk 84

4.5   Conclusions 84

Chapter - 5 LSE's Risk Based Profit Maximization under Severe Uncertainty .. 89

5.1   Introduction 89

5.2   Problem Description 90

5.3   Mathematical Model 91

    5.3.1   Procurement Cost 91

        5.3.1.1   Energy Storage System 91

    5.3.2   Energy Balance Constraint 92

    5.3.3   Revenue 92

    5.3.4   Sale Price Modeling 92

    5.3.5   Objective Function 93

5.4   Information Gap Decision Theory Method 93

5.5   Load Serving Entity's Decision Making: Information Gap Decision Theory Framework 95

    5.5.1   Decision Variables 95

    5.5.2   Price Uncertainty Modeling 95

    5.5.3   Performance Requirement 96

        5.5.3.1   Robustness Function 96

        5.5.3.2   Opportunity Function 97

5.6   Case Study 98

    5.6.1   Data 98

    5.6.2   Simulations 99

    5.6.3   Results 100

        5.6.3.1   Robust Strategy 100

        5.6.3.2   Opportunistic Strategy 103

        5.6.3.3   Sale Prices and Demand 104

        5.6.3.4   Consumer Energy Cost 106

5.7   Conclusions 106

Chapter - 6 LSE's Risk Based Profit Maximization Framework for Correlated Uncertainties 111

6.1   Introduction 111

6.2   Problem Description 112

6.3   Mathematical Model 112

    6.3.1   Procurement Cost 113

    6.3.2   Energy Balance Constraint 113

xiv

6.3.3   Revenue.. 113

6.3.4   Sale Price Modeling.. 113

6.3.5   Objective Function.. 114

6.4   Load Serving Entity's Decision Making: Information Gap Decision Theory Framework 114

6.4.1   Price and Demand Uncertainty Modeling............................ 114

6.4.2   Performance Requirement 115

6.4.2.1   Robustness Function... 116

6.4.2.2   Opportunity Function 118

6.5   Case Study... 119

6.5.1   Data 119

6.5.2   Statistical Calculations... 119

6.5.3   Simulations 120

6.5.4   Results.. 121

6.5.4.1   Robust Strategy.. 122

6.5.4.2   Opportunistic Strategy 122

6.5.4.3   Impact of Demand Uncertainty 125

6.6   Conclusions 126

Chapter - 7 Conclusions and Recommendations for Future Work 129

7.1   Introduction 129

7.2   Summary of Significant Findings 130

7.3   Recommendation for Future Work....................................... 132

REFERENCES... 137

PUBLICATIONS FROM THE WORK.. 153

APPENDIX-A 157

LIST OF TABLES

Table 3.1: Specifications of thermal self-generation unit 52

Table 3.2: Execution and computational time 54

Table 3.3: Expected profit under nominal and optimal dynamic sale price 61

Table 5.1: Specification of battery energy storage system 98

Table 5.2: Energy procurement from various sources 100

Table 5.3: Correlation between demand and spot market prices 106

Table 5.4: Consumer cost under flat and dynamic pricing 106

Table 6.1: Energy procurement from various sources 121

LIST OF FIGURES

Fig. 1.1: Thesis organization.. 12

Fig. 2.1: Major entities in restructured power supply systems 16

Fig. 2.2: LSE's operation in electricity markets 19

Fig. 2.3: LSE's decision making problem 21

Fig. 2.4: Risk reward mapping of retail prices [15]... 37

Fig. 2.5: Identified research challenges 39

Fig. 3.1: LSE's decision making.. 45

Fig. 3.2: Conditional Value at Risk 50

Fig. 3.3: Consumer classes' demand profile... 53

Fig. 3.4: Hourly spot market price and corresponding CVaR 53

Fig. 3.5: LSE sale price for consumer classes (a). Residential, (b). Agricultural, and (c). Industrial 56

Fig. 3.6: LSE demand for consumer classes (a). Residential, (b). Agricultural, and (c). Industrial 57

Fig. 3.7: Purchase from various procurement options for consumer classes (a). Residential, (b). Agricultural, and (c). Industrial.. 59

Fig. 3.8: LSE's revenue and cost for three consumer classes..................................... 60

Fig. 3.9: Efficient frontier for three consumer classes... 60

Fig. 3.10: LSE sale price for consumer classes (a). Residential, (b). Agricultural, and (c). Industrial 62

Fig. 3.11: LSE demand for consumer classes (a). Residential, (b). Agricultural, and (c). Industrial 63

Fig. 3.12: Purchase from various procurement options for consumer classes (a). Residential, (b). Agricultural, and (c). Industrial.. 65

Fig. 3.13: Efficient frontier for various consumer classes 66

Fig. 4.1: Bi-level decision making problem... 73

Fig. 4.2: PV generation profile 80

Fig. 4.3: Consumer demand profile 80

Fig. 4.4: Dynamics of sale prices.. 82

Fig. 4.5: Dynamics of consumers' optimized demand.. 82

Fig. 4.6: Procurement from various options for risk-neutral LSE 83

xvii

Fig. 4.7: Procurement from various options for risk-averse LSE 83

Fig. 4.8: Efficient frontier 84

Fig. 5.1: Illustration of information gap uncertainty 94

Fig. 5.2: Spot market price and demand profile 98

Fig. 5.3: Renewable profile 99

Fig. 5.4: Risk-neutral LSE's sale price and revised demand (a) with RE and BES and (b) without RE and BES 101

Fig. 5.5: Robustness and opportuneness curves for critical and anticipated target.... 102

Fig. 5.6: Energy procurement from different sources of LSE with RE and BES for (a) Robustness and (b) Opportuneness 102

Fig. 5.7: Energy procurement from different sources of LSE without RE and BES for (a) Robustness and (b) Opportuneness..103

Fig. 5.8: Sale prices and demand of LSE with RE and BES for (a) Robustness and (b) Opportuneness 104

Fig. 5.9: Sale prices and demand of LSE without RE and BES for (a) Robustness and (b) Opportuneness 105

Fig. 6.1: Pictorial representation of Ellipsoid bound info-gap model 116

Fig. 6.2: Demand and spot market price curve for weekday (Mon.) and weekend (Sun.) 119

Fig. 6.3: Correlation between spot market price and demand for weekday (Mon.) and weekend (Sun.)..120

Fig. 6.4: Sale prices for risk-neutral LSE..121

Fig. 6.5: Robustness and opportuneness curves for critical and anticipated target.... 122

Fig. 6.6: Energy procurement from various sources for (a) Weekday and (b) Weekend 123

Fig. 6.7: Expected cost for weekend and weekday 124

Fig. 6.8: Expected profit for various targeted profit for critical and anticipated target 124

Fig. 6.9: Hourly sale price for (a) Weekdays and (b) Weekends 125

Fig. 6.10: Impact of demand uncertainty on robustness and opportuneness curves .. 126

Fig. A.1: Cost curve of generating units 157

xviii

LIST OF ABBREVIATIONS

Abbreviation	Explanation
ARIMA	Auto Regressive Integrated Moving Average
B&B	Branch and Bound
BES	Battery Energy Storage
CDD	Conditional Drawdown
CPP	Critical Peak Pricing
CVaR	Conditional Value at Risk
DisCo	Distribution Company
DLC	Direct Load Control
DNO	Distribution Network Operator
DR	Demand Response
DRC	Downside Risk Constraints
DRP	Demand Response Program
DSO	Distribution System Operator
EDR	Expected Downside Risk
EDRP	Emergency Demand Response Program
ESS	Energy Storage Systems
EV	Electric Vehicle
GAMS	General Algebraic Modeling System
GB	Great Britain
GenCos	Generation Companies
IBDRP	Incentive Based Demand Response Program
IGDT	Information Gap Decision Theory
KKT	Karush-Kuhn-Tucker
LSE	Load Serving Entity
MCS	Monte Carlo Simulation
MF	Membership Functions
MILP	Mixed Integer Linear Programming
MINLP	Mixed Integer Non-linear Programming
MO	Market Operator
MSF	Market Share Function

Abbreviation	Explanation
PBDRP	Price Based Demand Response Program
PDFs	Probability Distribution Functions
PEM	Point Estimate Method
PJM	Pennsylvania-New Jersey-Maryland
PQC	Price Quota Curve
PTR	Peak Time Rebates
PV	Photovoltaic
RAROC	Risk Adjusted Recovery on Capital
RE	Renewable Energy
RES	RE Sources
RO	Robust Optimization
RPO/ RPS	Renewable Portfolio Obligation/ Renewable Portfolio Standards
RTP	Real Time Pricing
SBA	Scenario Based Analysis
SO	System Operator
ToU	Time of Use
TRANSCO	Transmission Company
TSO	Transmission System Operator
VaR	Value at Risk
VPP	Variable Peak Pricing
WEM	Wholesale Electricity Market

Chapter – 1: Introduction

This chapter briefly describes the background, context, motivation, and main contribution of the proposed work. It also provides structure of the thesis.

Chapter - 1

Introduction

1.1 Background

Load Serving Entity (LSE) is a new player in the electricity market, emerging due to deregulation and restructuring of power supply systems. It supplies electricity to the consumers that do not participate directly in the Wholesale Electricity Market (WEM). It performs the role of intermediator between consumers and generators/ producers of electricity by procuring electricity from WEM and selling it to consumers [1], [2]. Electricity procurement incurs cost and electricity selling generates revenue. The difference between revenue and cost is LSE's profit. This profit margin is generally narrow [3]. LSE maximizes this difference by minimizing procurement cost and maximizing revenue. Hence, LSE adopts a decision making process for optimal procurement and sale price decisions with an aim to minimize cost and maximize revenue.

Decision making is a process to make thoughtful, informed decisions. LSE makes procurement and sale price decisions for the medium-term horizon as it plans well ahead of actual electricity supply. Procurement decisions are made to determine optimal procurement portfolio, i.e., quantum of electricity to be procured from various available sources. Bilateral contracts, self-generation, and spot market are sources available to LSE in the medium term [1], [4], [5]. Prices of bilateral contracts and self-generation are fixed. However, prices in spot market exhibit variability throughout the day, i.e. high during peak hours and relatively low during off peak hours [6]. This characteristic of spot market prices leads to variability in procurement cost and thereby profit. This necessitates design of dynamic sale prices, which vary during different times of day, minimizing variability of cost and profit for LSE. On the consumer side, consumer demand varies through out the day, which may create mismatch between electricity to be procured and supplied by LSE. The excess/ deficit demand needs to be settled close to real-time [7]. This affects medium-term procurement decisions made to minimize cost and thus maximize profit for LSE. Hence, LSE faces variability of spot market prices and demand for advanced planning to procure electricity.

Consumer behavior influences demand to be fulfilled by LSE. Consumers exhibit an elastic behavior to the offered sale price, i.e., demand is a function of sale prices [7], [8]. Consumers respond to change in sale prices by modifying their demand. The change in demand depends on consumers' price sensitivity or elastic demand. Therefore, changes in offered sale prices influence price sensitive consumers' electricity consumption. Depending on the level of competition, retail competition is another factor that may influence sale price setting of LSE [1], [9].

Retail competition provides opportunities for consumers to select their supplier (LSE) based on price and service quality [10], [11]. Consumers may switch to LSE offering lower prices [12]. Consumer switching may reduce LSE's market share and affect its profit. Low sale prices might be unable to recover electricity procurement cost that leads to losses for LSE. Level of retail competition depends on consumer switching rate. Globally, consumers are not very disposed to change suppliers, and incumbents are not challenged by competition from new entrants. Consumers switching rate is low (below 10%) in countries like Italy, Denmark, France, Germany, the Netherland and Belgium. The causes for low switching rate are switching barriers such as high search cost, high switching cost, and limited information provision. Hence, a large proportion of consumers never switch their supplier, even though significant discounts are available [13]. This implies low switching or weak retail competition. Though consumers willing to reduce their electricity cost prefer to shift their consumption from one hour to another. Hence, consumer behavior plays a vital role while making decisions after evaluating sale prices.

LSE offers electricity under flat rate or time varying/ dynamic prices to consumers [1], [8]. Flat rate price remains fixed and do not vary with time. These prices do not reflect actual supply cost. Consumers do not face the varying cost of power supply with a change in demand. Hence, they do not face challenges of high electricity bills for any unplanned electricity consumption. On the other hand, LSE has to procure this unplanned quantum of electricity at varying prices, which may increase its procurement cost. Under time varying/ dynamic pricing, price depends on time of electricity consumption, i.e., price varies from one time or block to another time or block [14]. These prices reflect expected or observed changes in supply and demand over time and their impact on costs. Time varying/ dynamic tariff structures have the potential to modify consumer's demand profiles [15]. Also, time varying/ dynamic price reduces exposure to price risk during peak hours of wholesale electricity market for LSE. Such pricing provides an opportunity for consumers to reduce their electricity bill by modifying their demand based on price signals [12], [16]. Therefore, setting of dynamics of sale price is important while LSE makes decisions to maximize profit.

In a liberalized electricity market, LSE has no control over electricity prices in spot market as competitive forces are responsible for its determination [6]. Prices reflect system demand-supply balance. Therefore, prices are no more stable but characterized by volatility and uncertainty. Non-storability of electricity, mismatch between supply and demand, demand forecast error, requirement of peaking plant during peak hours, generator outages, varying weather conditions, and other unknown market forces make electricity prices highly volatile than any other commodity. This price volatility can be more than 300% as compared to other commodities [17]. Consumer demand is another uncertain parameter in LSE's decision making [7], [18]. As future electricity prices and demand are unknown to LSE (decision-makers) at the time of planning, it experiences uncertainty in evaluating outcomes from electricity trading. Therefore, considering the level of volatility and uncertainty, price and demand are major threats to LSE and pose financial risk to substantially affect its profit.

Risk indicates possibility of harm or loss arising due to unfavorable outcomes of decisions. Possibility for adverse outcomes can be mitigated by designing appropriate risk management strategies. A typical risk management process includes identification, assessment, and control of risk. Hence, it fundamentally necessitates prudent decision making to design risk management strategies to secure profit and control possibility of loss or risk.

1.2   Motivation for the Proposed Work

Evolving competition in the electricity markets comes up with opportunities and challenges for LSEs. Concerns of effective and strategic decision making have been raised to exploit benefits from opportunities while managing/ handling associated challenges. This is driving enormous interest of research community towards strategic decision making.

Electricity market structure plays a significant role in LSE's decision making problem. Presently, in many countries, wholesale market is competitive, but retail markets are

Chapter - 1

not exposed to competition due to fear of consumer exposition to high prices and political influences. Therefore, non-existence or weak competition in the retail market poses different challenges than what is raised by the perfectly competitive retail market. Consumer characteristics/ behavior are another factor that potentially changes the nature of LSE's problem [12]. Consumers' characteristics vary from one consumer to another [19]. Considering this, LSE can offer electricity under flat rate or dynamic prices.

Flat rate is a conventional pricing scheme adopted in most of the markets [7], [20]. As flat rate represents average of the procurement cost, it cannot reflect cost difference of supplying electricity at different time periods. Also, lack of incentives in flat rate do not encourage consumers to modify/ adjust their electricity consumption [20]. Therefore, potential benefit of consumer response from demand shifting cannot be utilized. These issues with flat rate restrict LSE to advantage from low spot market price hours, and to maximize its profit.

Dynamic prices provide opportunity to offer cost reflective prices [21]. Also, it enables LSE and consumers to exploit benefits of Demand Response (DR). DR is a strategy used to modify consumers' electricity consumption from its normal consumption profile [22]. LSE tries to shift electricity consumption from peak hours to leaner demand hours to get benefit during low price hours of spot market. This requires reflection of consumer's price responsive behavior, i.e., how consumer respond to offered sale prices in setting dynamic prices. Price-sensitivity drives consumers to respond to offered sale prices [12]. Therefore, sale prices must be set to adequately incentivize consumers according to their price sensitivity to attain flexibility. Though demand profile and price sensitivity vary from consumer to consumer, they can be categorized into different classes. Each consumer class is required to be treated differently for sale prices and procurement decisions [20]. This necessitates consideration and analysis of different consumer classes in LSE's decision making.

Dynamic sale prices modify consumer demand profile [19], [23]. Therefore, it impacts quantum of electricity to be procured, and consequently procurement portfolio decisions. This implies that these decisions are interrelated with each other in one or another way. However, traditionally, LSE's procurement and sale price determination problem are solved separately. When energy procurement and sale prices decision problem are solved jointly, it can enhance LSE's profit [24].

LSE procures electricity from power exchange, bilaterally from producers, and can schedule dispatchable self-generation to supply a part of demand. Self-generation is considered during high price hours to reduce LSE's procurement cost [25]. However, Renewable Energy (RE) based self-generation facilities need to be scheduled based on their availability due to its non-dispatchability. Available RE is required to be scheduled before scheduling of electricity from spot, bilateral, and thermal-based self-generation. RE characteristics impact procurement and sale price decisions [26]. LSE's procurement and sale price decisions require investigation targeting RE availability.

As electricity prices are volatile and uncertain, they pose price risk for LSE. This risk needs to be minimized to get expected/ targeted profit [27]. Therefore, more stress on uncertainty modeling and risk management is required. Risk is quantified using risk measures. Mean-variance, Value at Risk (VaR), and Conditional Value at Risk (CVaR) are some of popular risk measures [6]. As CVaR is coherent, it has become a popular and well accepted risk measure. However, CVaR, like other risk measures, requires specific information of distribution function of uncertain parameters to quantify risk [28]. Though, with lack of information about uncertain parameters, distribution function cannot be determined. Under such situations, CVaR cannot be considered to model uncertainty of parameter of interest [29]. Further, considering that uncertainty can have adverse and favorable faces, it may lead to loss and benefits to the decision maker, respectively [29]. As CVaR is confined to tail risk or loss, it cannot help decision makers to make decisions when uncertainty has favorable face. This necessitates consideration of an alternative approach for risk management with adverse and favorable uncertainty.

Information Gap Decision Theory (IGDT) is widely adopted by finance and power system community to make decisions under adverse and favorable uncertainty [30]. Unlike other uncertainty modeling approaches, IGDT requires less information about parameter of interest for uncertainty modeling. IGDT models uncertainty without the knowledge of underlying probability distribution function. IGDT exploits available information of uncertain parameters and informs decision maker about the negative and positive outcomes that can be caused by uncertainty to take appropriate and logical decisions.

LSE faces uncertain spot market price and demand while it makes risk based decisions to maximize profit. These uncertain parameters may exhibit certain relationships.

Chapter - 1

Statistical analysis on spot market prices and demand highlights their correlated nature [6], [31]. This correlation between spot market price and demand may be strong or weak depending on the size of LSE [2], [31]. LSE with low demand indicates weak correlation and vice-versa. This correlation highlights possibility of extreme situations, which may cause substantial financial losses to LSE. Consideration of correlated variations of price and demand can help to secure LSE's position in the electricity market and may improve its profit margin. This requires a single framework to model correlation between uncertain parameters. IGDT has the capability to model dependence between multiple uncertainties in a single framework. Ellipsoid bound model can capture correlation between price and demand. The shape and orientation of uncertainty-ellipsoids are defined by the variance-covariance matrix. Based on this, uncertainty is quantified, and decisions are made.

1.3   Contribution of the Proposed Work

This research work aims to support LSE to make prudent procurement and dynamic sale price decisions. These decisions are determined for various decision making models of LSE to maximize its profit. Research objectives include determination of dynamic sale prices and procurement portfolio targeting spot market price uncertainty, consumer's price responsive behavior, and renewable energy availability. Considering importance of correlation between spot market price and demand, a novel framework is presented for LSE's decision making. The summary of main contributions of the proposed work are as follows

- Traditionally, LSE's electricity procurement and sale price decision making problem were modeled and solved separately. However, these are interrelated with each other. Hence, these problems are jointly modeled and addressed in this thesis. Consumer's elastic demand is modeled to set dynamic sale prices and to determine energy procurement decisions. These decisions are analyzed for various consumer classes.
- Considering spot market price uncertainty as most volatile than any other uncertainty involved in LSE's decision making, price uncertainty is modeled, and risk is minimized by CVaR. A number of strategies are determined for different risk aversion levels and are presented by an efficient frontier. These risk-based strategies are analyzed.

- RE availability significantly impacts LSE's sale prices and procurement decisions. Hence, the proposed model is extended to analyze impact of renewable availability on LSE's decisions. Cost of renewables is considered to reflect economic aspect of LSE's decision making.

- Consumer determines consumption pattern rationally instead of its price elasticity to minimize energy bill. Hence, consumer has an economic objective to minimize cost. This consumer's objective is conflicting to LSE's objective. These objectives are modeled as interaction between LSE and consumer to enhance their economic benefits. For this, an advanced mathematical programming: bi-level programming-based model is considered.

- As future prices are influenced by various unknown and unpredictable factors, accurate and error-free forecasting of prices is not possible. Lack of information about price leads to a model based on assumptions. Therefore, reliability of decisions taken cannot be guaranteed. IGDT framework is considered to obtain robust decisions. This can provide reliable decisions with available information of uncertain parameter: price. This framework also helps to decide decisions under favorable face of uncertainty. Benefit of ESS scheduling is exploited to enhance LSE's profit and reduce risk.

- Though price and demand uncertainties are considered in LSE's decision making, correlation between them, which is an important consideration, is not considered in single framework. As this correlation is strong for a large LSE, co-variations in price and demand would significantly influence LSE's profit margin. This thesis contributes by modeling correlated spot market price and demand in a single framework for a large LSE. A novel framework based on IGDT is presented to highlight importance of correlation in LSE's decision making. Ellipsoid bound info-gap model is considered to model the correlation.

This thesis contributes by presenting various risk based decision making models for an LSE to maximize its profit. These models consider consumer price responsive behavior, RE availability, and price uncertainty, which significantly affects LSE's profit. Correlated price and demand uncertainty is modeled to highlight their impact on LSE's decisions.

Chapter - 1

1.4 Scope and Limitation

Scope of work is to develop decision making models suitable for LSE operating in competitive electricity markets. As the level of competition in retail market is different for countries exposed to full competition, knowledge of competition is a must. The weak retail competition or low consumer switching hardly impacts incumbent LSE's market share. However, decisions about sale prices and procurement are significantly affected by competition at wholesale market. Offered sale prices can drive price sensitive consumers. Therefore, benefits of demand shifting are exploited to devise best strategy for sale and procurement of electricity to maximize profit. The decision making model presented in this thesis is suitable for a large LSE operating in weak retail market.

1.5 Thesis Organization

This thesis is organized into seven chapters. Connectivity between chapters and chapter outline are shown in Fig.1.1. Theoretical background of problem and proposed research objectives are indicated.

Chapter 1 presents a general background of LSE's decision making. It explains the motivation behind the proposed work, major contribution, scope and limitation, and organization of the thesis.

Chapter 2 provides a comprehensive review of the literature associated with LSE's decision making. Consumer behavior concerning offered sale price and its modeling in sale price determination is discussed. Uncertainty handling techniques and risk management approaches adopted in LSE's decision making are discussed. Discussion on uncertainty is extended to highlight importance of modeling correlated price and demand uncertainties. Based on this review, identified research challenges are presented at the end of the chapter.

Chapter 3 presents a decision making model to determine sale prices and procurement for elastic consumer demand. Both procurement decisions and sale prices are determined simultaneously, in contrast, to solve these problems separately. This base model is extended to include RE availability to analyses its impact on sale prices and procurement decisions. Risk-neutral and risk-averse behavior of LSE for cases with and without RE are discussed.

Introduction

Chapter 4 presents a mathematical model to capture interaction between LSE and consumer. Considering that consumer would behave rationally rather than based on elasticity, its economic objective of cost minimization is formulated. Using bi-level programming, LSE's and consumer's conflicting objectives are formulated in two-level. Impact of RE availability on LSE's decisions under this framework are analyzed. Obtained dynamic sale prices and procurement decisions are discussed for risk-neutral and risk-averse LSE.

Chapter 5 presents IGDT based risk management approach to model uncertainty. It supports situations where sufficient information is not available about uncertain parameters. LSE's risk-averse and risk-seeking behavior to achieve targeted profit is modeled by robustness and opportuneness functions. In this model, ESS is considered in addition to WEM and RE to analyze its impact on LSE's decision making. Results obtained for risk-averse and risk-seeking cases are investigated and discussed.

Chapter 6 describes correlated spot market price and demand modeling under IGDT framework. Ellipsoid bound model is selected to capture correlation between spot market price and demand. Both risk-averse and risk-seeking behavior of LSE are modeled. The proposed model is illustrated via a case study conducted for weekdays and weekends, highlighting correlation. Impact of demand uncertainty on LSE's decisions is highlighted.

Chapter 7 presents overall summary, conclusion and discusses future research direction.

Chapter - 1

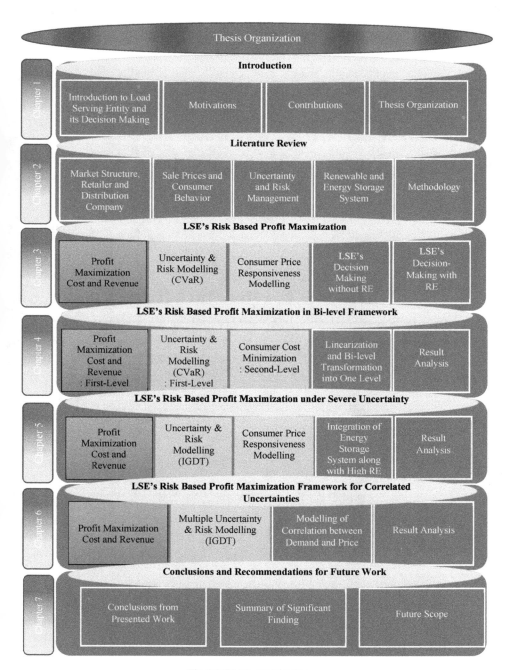

Fig. 1.1: Thesis organization

Chapter – 2: Literature Review

This chapter presents a state-of-the-art review of LSE's decision making.

Chapter - 2
Literature Review

LSEs face various challenges in evolving power supply systems. Increasing share of renewables, consumer behavior, and uncertain electricity prices and consumer demand are challenges that significantly affect LSE's profit. These challenges call for novel decision making models to maximize LSE's profit. This chapter presents a state-of-the-art review of LSE's decision making. Subsequent sections provide detailed discussions about industry structure, procurement strategy, pricing strategy, market uncertainties, consumer behavior, renewable energy, and energy storage considerations for LSE's decision making.

2.1 Evolving Industry Structure

In a deregulated paradigm of electricity supply systems, the industry's structure is evolving to enhance efficiency, improve supply quality, maximize consumer choices, and promote competition. In the evolving industry structure, generation, transmission, distribution, and retail businesses operate independently [2], [32] and the role of traditional utilities is redefined. New entities/ players are emerging to perform their role in power supply business and power markets [33]. Generation Companies (GenCos) and LSEs are responsible for generation and supply of electricity, respectively. System Operator/ Transmission System Operator (SO/ TSO), Distribution System Operator/ Distribution Network Operator (DSO/ DNO), and Market Operator (MO) play significant roles in the effective and efficient operation of power supply systems and electricity markets [34]. These entities strategize to support physical and economic operations of power supply business.

Fig. 2.1 shows the structure of a restructured power supply system. LSE procures electricity from generator, i.e., GenCo and supplies to consumers, and utilizes transmission and distribution networks to transfer that power. Transmission Company (TRANSCO) owns transmission network and provides networks for electricity transmission. SO or TSO manages the system for load generation balancing. Distribution Company (DisCo) or DNO operates distribution network. Consumers are end-users of electricity and may procure electricity from LSE or WEM depending on the quantum of electricity required. Generally, consumers with large demand (>1 MW) with connectivity to transmission network can participate in WEM. Small consumers, i.e., consumers with demand <1 MW procure electricity from electricity suppliers; LSEs. These consumers can be classified into various classes such as residential, agricultural, and industrial based on their engagement types [33], [35]. The flow of electricity, information, and money among various entities in restructured power systems is indicated in Fig. 2.1.

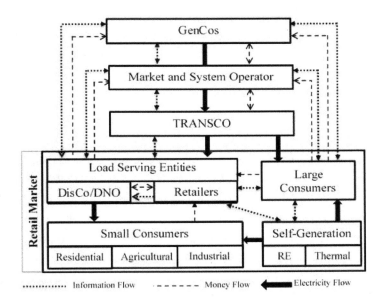

Fig. 2.1: Major entities in restructured power supply systems

Benefits of liberalizing the electricity sector and having a competitive WEM can be transferred to consumers with the retail markets [13]. In retail markets, LSEs compete for consumer demand by offering attractive sale prices. LSEs' strategies are specifically targeted to maximize their profit, i.e., governed by the economics of buy and sale of electricity [1]. As LSEs have identical sourcing costs (competitive WEM), they have no choice but to keep their profit margin as optimal as possible. LSEs are expected to pass benefits to consumers when WEM prices fall to maintain or enhance their market share [13].

Retail pricing plays a significant role in retail market dynamics and affects competition among LSEs. Fixed pricing schemes have been considered to be a limitation for LSE's profit maximization [1]. Flexible pricing schemes, particularly, targeting consumers' price responsive behavior are required for electricity retailing. This motivates the development of effective decision making models to maximize LSE's profit. An extensive focus on retail pricing and consumer price responsive behavior in the model would devise effective market strategies. Therefore, this thesis focuses on LSE's profit maximization strategy in a liberalized electricity market.

2.2   Load Serving Entity

LSE's primary role is to procure electricity on behalf of consumers from WEM and supply it to them at contracted/ predefined sale prices under various pricing schemes [1]. Therefore, LSE is an intermediator between WEM and consumers, and its core business is to procure and sell electricity.

Traditionally, DisCos, who own and operate distribution network, were the electricity suppliers to consumers and performed the role of LSE. In the liberalized electricity supply systems, retailers are advented to supply electricity to consumers [2]. LSE is a newly advented retailer and is also known as power marketer [2], [6]. Retailers do not own any physical assets or position in the electricity supply systems, unlike DisCo. Retailers utilize distribution network services of DisCo to channelize electricity to consumers. Retailers do not care about technical challenges such as congestion, and service interruption due to line outages and line overloading, etc., and operating cost of network [24]. Therefore, relationship of retailer with DisCo is limited to the payment of network access tariff associated with each consumer.

Several countries are progressively experiencing system restructuring. Retail segment is evolving to restructuring and exposed to full/ partial retail competition. In such cases, incumbent suppliers or DisCo supply electricity and are generally regulated [23], [24]. Therefore, retailers and DisCo can be termed as LSE. In this thesis, term "LSE" represents retailer's or DisCo's market operation (trading) only, i.e., network operation is neglected to highlight their market strategies required for profit maximization.

2.3   Load Serving Entity's Operation in Electricity Markets

LSE's operations in the electricity markets are divided into supply side and demand side. It sets up contracts for procuring and selling electricity for supply and demand sides, respectively [2], [18]. In supply side contracts, its objective is to procure

Chapter - 2

electricity at the minimum possible cost from WEM. At the demand side, it sells electricity at optimal sale prices to maximize revenue. Difference between revenue and electricity procurement cost is LSE's profit. Therefore, LSE aims to maximize profit by minimizing procurement cost by strategically participating in WEM and maximizing revenue by offering optimal sale prices to its consumers. Subsequent sub-sections briefly describe the supply and demand side operations.

2.3.1 Supply Side Operation

On the supply side, a significant challenge for LSE is to accurately determine the procurement portfolio [1]. Optimal quantum of electricity procurement from multiple sources of WEM forms procurement portfolio, which minimizes LSE's procurement cost [36]. For this, LSE identifies available energy sources (Fig. 2.2). LSE may sign fixed power and fixed price bilateral contracts with GenCos, or may also participate in forward and spot markets [5], [18]. For transactions under bilateral contracts, negotiation on terms and conditions such as price, delivery points, and transmission charges is a challenge for LSE and GenCo. Forward contracts are similar to bilateral contracts, contracted for buying or selling of fixed quantum of electricity at a predefined rate for future delivery date. In contrast to bilateral contracts, forward contracts have standardized terms and conditions. Forward contracting curve defines the prices at which a contract (for certain quantum of power) for future delivery can be concluded [18], [33]. These curves are supposed to be known at the time of signing forward contracts. Self-generation based on thermal and renewable resources may be utilized for electricity procurement [25], [37], [38]. Emerging Energy Storage Systems (ESS) such as hydrogen, pumped hydro, and battery-based storage systems are considered for energy scheduling [39]–[42]. Depending upon market types, LSE may have choices to select future contracts for electricity procurement. Call option is a common future contract that gives the right to purchase fixed quantum of power at a predefined strike price before the option reaches its expiration date. On the request of LSE, sellers of call option exercise contracts for which they charge some premium(fee). A premium is paid for securing the right to exercise or non-exercise call option [43].

Uncertainty in electricity markets, i.e., price and demand uncertainty influences LSE's trading decisions [3]. These uncertainties pose significant risk to LSE [6], [34]. Risk is the possibility of harm, threat, loss, or any other negative occurrence that is caused by external or internal vulnerabilities [27]. This risk may result in losses that reduce LSE's profit. Therefore, LSE tries to minimize risk. Hence, LSE has two objectives: profit maximization and risk minimization [31], [44]. These objectives are conflicting or opposite, for which LSE determines optimal trading strategies.

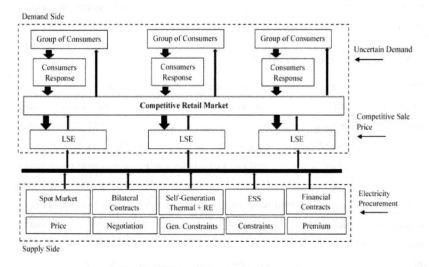

Fig. 2.2: LSE's operation in electricity markets

Bilateral, forward, and financial contracts are considered to minimize the risk exposure of spot market [34], [43], [45]. Risk aversion term is the risk tolerance parameter or weighting factor that prevents LSE from trading large quantities in highly volatile spot market [7]. Therefore, it reduces LSE's risk exposure to volatile prices in spot market. However, this consideration might decrease LSE's expected profit. Desired level of risk exposure is defined by value of weighting factor [6], [7]. A value of zero indicates that LSE is a risk-neutral or risk-seeker. A value higher than zero indicates that LSE is risk-averse [43]. Different values of risk weighting factors help to derive efficient frontier. Efficient frontier indicates possible trade-off between expected profit and risk of profit volatility, for different risk weighting factor. Risk of profit volatility, i.e., standard deviation of profit is quantified using risk measures [6]. Selection of its value depends on the financial situation of LSE and its willingness to take risks, and market characteristics [7], [23]. A detailed discussion on uncertainty and risk handling approaches is presented in subsequent section.

Chapter - 2

2.3.2 Demand Side Operation

The upper portion of Fig. 2.1 indicates demand side operation of LSE. On the demand side, sale price determination is affected by consumer behavior and retail competition [32], [46], [47]. Consumers exhibit certain behavior/ characteristics regarding electricity consumption concerning offered sale prices. Generally, consumers respond to offered sale prices by modifying their consumption based on their willingness to pay or price sensitivity to change in offered sale prices [6], [20]. Such consumer behavior is characterized as elastic or price responsive behavior. Willingness to pay indicates the maximum price at or below which a consumer will definitely purchase one unit of product. Alternatively, it is defined how high a consumer is willing to go when purchasing a unit of product. Whereas, price sensitivity indicates the extent to which consumer demand changes with sale price changes [8]. Price sensitive consumers redistribute their demand considering change in sale prices to minimize cost, without reducing their overall demand. Hence, sale prices potentially affect consumer demand and consumption pattern. Therefore, consumer behavior modeling, i.e., demand-price relationship modeling is an important aspect while sale price is decided by LSE.

Retail competition provides opportunities to consumers for choices of LSEs and various contracts/products offered by them. Generally, consumer selects or makes a choice of LSE based on offered sale prices [48]. Low sale prices may attract consumers, but may not recover cost associated with electricity procurement and generate adequate revenue. For high sale prices, consumers could switch to another LSE [18]. Therefore, offered **sale prices of rival LSEs may affect LSE's market share. As retail market is all about** price competition, level of retail competition will significantly influence prices offered by LSEs. Consumer switching rate primarily indicates level of retail competition. A high consumer switching rate signifies strong retail competition and vice-versa.

Consumer switching depends upon whether a large fraction of consumers is active or passive. Many passive or dormant consumers and a lack of active consumers' participation indicates that they do not search or switch, rather stay with the default LSE. For example, currently, more than 50 percent of the consumers in Great Britain (GB) are still on a default tariff [13]. Also, consumers are reluctant to choose new suppliers (LSE) due to risk of high electricity bills. Other reasons include uncertainty about service quality of new suppliers, high switching cost, complexity of retail market and tariff plan, and low awareness about pricing and products. Lack of consumer engagement has been one of the biggest weaknesses of liberalized retail electricity markets. This is responsible for low switching rate and hence indicates weak retail competition. Under such conditions, retail markets remain concentrated with aggregated market share for the incumbent LSEs.

LSE's decision making problem is illustrated in Fig. 2.3. It is worth noting that LSE makes sale price and procurement decisions well in advance, i.e., a few days to a few months ago. Hence, its decision making falls under medium-term planning [6].

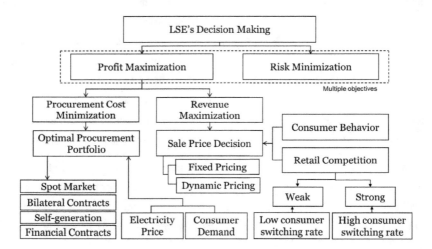

Fig. 2.3: LSE's decision making problem

2.4 Market Uncertainties

As LSE procures electricity from the spot market of WEM, any variation in spot market price impacts its procurement cost. Spot market price is dynamic, changing based on the demand for electricity and availability of supply. It is determined by market forces and depends on various known and unknown factors such as supply, demand, and bidding strategies of participants. This characterizes spot market price as volatile and uncertain [6]. Spot market price volatility is higher than any other commodity particularly due to its unique characteristics of non-storability [17]. California's electricity market has witnessed extreme volatility in hourly prices [17]. This market crisis resulted in huge financial loss for supplier companies (LSEs), and as a

Chapter - 2

consequence they went bankrupt. Therefore, spot market price volatility and uncertainty cannot be neglected when LSE's procurement decisions are being addressed.

LSE is obligated to supply electricity to its consumers upon request [43]. However, LSE cannot forecast the demand precisely as consumer demand fluctuates and is uncertain. Hence, in real-time, actual and procured demand could be different [43]. Difference in the quantum procured and supplied energy force LSE to settle this mismatch in spot market [2]. Hence, it leads to risk in procuring energy mismatch at prevailing spot market prices and impacts LSE's procurement cost. LSE aims to avoid losing its small margins by procuring at prevailing spot market prices. This requires an accurate estimation of demand obligation [2]. Due to lower volatility, demand uncertainty impacts minimally on LSE's profit when compared to spot market price uncertainty [43], [46]. Risk modeling of uncertain parameters is essential for LSE to manage profit and risk. Uncontrolled exposure to risks could lead to devastating financial consequences [45]. Further, spot market prices and demand are inherently correlated and may influence LSE's trading decisions depending on its nature and strength [2], [6]. This has been discussed in the subsequent subsection 2.4.1.

LSE faces uncertainty in consumer behavior with respect to sale price signals. Consumers reaction to electricity prices can be modeled by price elasticities. However, consumer reaction may vary due to factors such as income and comfort. Hence, LSE may face uncertainty in price elasticities when it determines sale prices [49], [50]. Modeling price elasticity uncertainty is essential when consumer behavior is uncertain or is continuously changing due to changes in technological or social phenomena [49]. Details on price elasticity are presented in the forthcoming Section 2.6.

2.4.1 Correlated Price and Demand

Correlation between demand and spot market prices in LSE's decision making are considered from time-to-time [2], [3]. Statistical analysis of demand and spot market prices highlight their correlated nature. However, strength of this correlation depends upon the LSE size. This correlation may be weak for an LSE with low demand, in particular for small retailers [3]. Under such conditions, use of separate marginal probability distribution for demand and prices is opposed and a joint probability distribution is recommended [2]. However, correlation between demand and spot market prices is usually strong for an LSE with large demand [31]. Correlation between demand and spot market price increases the probability of extreme situations, which may cause substantial financial losses to LSE. Nonetheless, correlated variations of demand and spot market prices can secure LSE's position in the electricity market and improve its profit margin [6]. Therefore, joint probability distribution is imperative for uncertainty modeling. However, determination of joint probability distribution is complex and cumbersome. Hence, this makes uncertainty modeling challenging.

Correlated demand and prices in LSE's decision making have been considered *via* a variety of ways. Correlation between demand and spot market prices has been modeled using scenario tree approach and time series model under a stochastic framework in LSE's decision making problem [43], [6]. However, scenario tree cannot consider extreme situations (spikes), which are pronounced in higher frequency at hourly level, and may lead to imprudent decisions. Further, this method suffers when strong correlation exists between modeled variables [31]. Dynamic regression is used to model correlation between demand and price [36]. For this, a set of scenarios is generated using Auto Regressive Integrated Moving Average (ARIMA) model. However, to forecast price in each period, it is necessary to know the demand in each period. Further, ARIMA model assumes that stochastic process is stationary, which is not necessarily true. Such issues with existing frameworks necessitate a novel framework to model demand and spot market price uncertainties, along with their correlation. Further investigation on this may help LSE to incorporate correlated demand and price uncertainty in single framework with ease.

2.4.2 Uncertainty Handling Approaches

Uncertainty characterization and its mathematical modeling is an essential part of any risk-based decision making process [6], [51]. Uncertainties can be modeled using deterministic, probabilistic approaches [2], [3], [7], possibilistic approaches [51], hybrid probabilistic and possibilistic approaches [51], stochastic programming [52], IGDT framework [26], [53], Robust Optimization (RO) [37], [54], and interval analysis [55]. Despite availability of these models, the choice of suitable model depends upon the type of uncertain input parameters and decision making problem under consideration.

Chapter - 2

2.4.2.1 Deterministic Models

Deterministic models focus on the use of point prediction, which provides a single value of expected uncertain parameter for a given lead-time. This value does not provide any information about the possible deviation from forecasted value. Solutions based on deterministic forecasts may not provide desired accuracy.

2.4.2.2 Probabilistic Techniques

Probabilistic techniques capture statistical properties of uncertain parameters using Probability Distribution Functions (PDFs) [51]. This technique can be classified as Monte Carlo Simulation (MCS), Scenario Based Analysis (SBA), and Point Estimate Method (PEM). MCS is the most widespread probabilistic technique used to estimate PDF of spot market prices and demand in LSE's decision making [46], [56], [57]. MCS can be implemented using different sampling technologies such as Latine hypercubes sampling and Markov chain [58]. However, MCS has poor computational efficiency and intractability due to a large number of scenario requirements. Moreover, probabilistic solutions are hard to comprehend; hence it is problematic to make decision based on such solutions.

2.4.2.3 Possibility Theory

Possibility theory provides an appropriate framework for the analysis of uncertainties with an epistemic nature. Uncertain input parameters are regarded as Fuzzy variables, characterized by specific Membership Functions (MF) [59]. Lack of well-defined criteria for the selection of suitable MF is a major drawback of this approach [58].

2.4.2.4 Hybrid Possibilistic-Probabilistic Techniques

When it is possible to characterize some of the uncertain parameters by possibilistic and some of them with probabilistic techniques, a hybrid possibilistic-probabilistic technique is the best choice to model uncertain parameters [51]. Solution will be based on collective impact of probabilistic and possibilistic uncertainties. However, this method is time consuming.

2.4.2.5 Stochastic Programming

In stochastic programming, modeling each uncertain parameter requires a set of finite outcomes, which represent plausible realization with corresponding probability of occurrences [60]. In this formulation, randomness of input (uncertain) parameters is described by stochastic processes. Possible outcomes of random inputs are represented by scenarios, which are used by the stochastic process. A large number of scenarios are required to capture complete characteristics of uncertain parameters. Scenario can be generated through a scenario generation technique such as scenario tree, roulette wheel mechanism [33], MCS [56], Moment matching or time-series model [43] over the planning horizon [24]. However, a large number of scenarios make the problem intractable and increase its computational burden [33]. Therefore, these scenarios are reduced to a small number to regain tractability. These reduced scenarios must carry all statistical properties of the original set of scenarios. Scenario reduction techniques such as fast forward selection or simultaneous backward reduction methods are used to obtain small number of scenarios. Clustering techniques such as the k-means algorithm are considered to obtain few clusters that possess original characteristics of large scenarios [36].

There are no certain conditions or specified rules that state the requirement of minimum number of scenarios to characterize uncertain parameters. Traditionally, scenarios with a probability less than a threshold are discarded to determine the number of scenarios. However, determining the value of threshold remains debatable. Moreover, scenarios can be reduced to a number to obtain certain trade-off between computation time and solution accuracy [33].

2.4.2.6 Information Gap Decision Theory

Uncertainty modeling with possibilistic and probabilistic approaches require pre-specified PDF or MF [61]. This can be derived efficiently if massive historical information/ data of uncertain parameters is available. However, due to lack of information about uncertain parameters, PDF or MF cannot be obtained [51], [62]. PDF or MF is determined based on certain assumptions, which may lead to imprudent decisions. In such cases, IGDT approach may be considered to model uncertainties [29], [30]. IGDT-based models are not sensitive to random variable forecasts, as they do not

Chapter - 2

require any probabilistic estimation of uncertain parameters. Moreover, this method requires relatively small information inputs as compared to alternative theories of uncertainties.

For the IGDT approach, uncertain parameters are approximated *via* variation intervals [29]. IGDT falls under performance satisfying instead of performance maximizing approaches. Optimal decisions are specified based on the desired performance (or acceptable profit threshold), which is defined by the LSE. IGDT assesses desired performance by robustness and opportuneness functions. These functions are called decision-immunity functions and address risk-averse or risk-seeking behavior of decision-makers [62]. This technique is considered to evaluate robust strategy under severe uncertainty [63]. Severe means a 'very large' uncertainty space and poor initial estimates of the uncertain parameters in this space [64].

Various info-gap models, Energy-bound, Minkowski-norm, Slope-bound, Fourier-bound, Hybrid info-gap, and so forth are available to characterize uncertain parameters with specific features [29]. Envelop bound info-gap model is one of the popular IGDT models, which is applied to various decision making problems of power systems [65]–[69]. IGDT can handle multiple uncertainties together to evaluate robust or opportunistic trading strategy [29]. Envelop bound info-gap model can be applied to characterize multiple uncertain parameters [63], [70], [71]. Such formulation results in multi-objective optimization and requires Pareto optimality concept to determine a single solution. However, envelop bound info-gap model cannot capture correlation between uncertain parameters, as each uncertain parameter is modeled by independent info-gap model. Ellipsoid bound info-gap model is applied to capture auto correlation and correlation between uncertain prices from various markets [72], [73].

2.4.2.7 Robust Optimization

Similar to IGDT, RO does not require PDF or MF, but uncertainty sets are essential. RO represents uncertain parameters by parametric bounds or interval range [74]. Typically, upper and lower limits of uncertain parameters determine uncertainty sets. This is followed by selection of representative scenarios from the uncertainty set. RO provides solution for the uncertainty set including worst-case scenario [59]. RO based solution remains feasible for all values of uncertainty data within uncertainty set. However, such solutions are conservative as these are based on worst-case scenario.

2.4.2.8 Interval Analysis

Interval analysis characterizes uncertain parameters based on a known interval. This method is similar to probabilistic modeling with a uniform PDF [40], [51]. This method can be used when just an interval exists for uncertain parameters.

2.5 Risk Management Approaches

Exposure to uncertainties leads to risks, which necessitates LSE to manage risks by appropriate risk management approaches. For this, it is prevalent to perform risk assessment for energy trading of LSE [27]. Risk could be assessed by a risk measure such as variance [56], Value at Risk (VaR), and CVaR [6], [23], [28].

Variance is the most straight forward way to measure risk. It is applicable when uncertainty follows normal or log-normal distribution [75]. Variance fails to capture asymmetry of distribution and fat tails [76]. VaR is tail-based risk measure that summarizes the expected maximum loss over a targeted horizon for a specific confidence interval. VaR does not satisfy coherence property which is a serious drawback. Hence, it may not effectively capture portfolio diversification advantages [36].

CVaR is a percentile and coherent risk measure, which overcomes limitations of VaR. It satisfies properties of monotonicity, sub-additivity, homogeneity, and translational invariance [75]. It provides a better quantification of potential losses exceeding the specified confidence level. For a confidence level α', CVaR is defined as the expected profit of the $(1-\alpha')*100\%$ scenario with low profit [75]. CVaR is also called mean excess loss, mean shortfall, or tail VaR. Considering the potential of CVaR, it is widely used for risk management in various decision making problems [9].

Other risk measures such as Risk Adjusted Recovery on Capital (RAROC) [5], Expected Downside Risk (EDR), Conditional Drawdown (CDD), and Downside Risk Constraints (DRC) have also been used to quantify risk. RAROC determines a ratio between expected investment return and economic capital to quantify the risk [77]. This risk measure is considered to quantify risk associated with WEM and bilateral contracts procurement for an LSE. EDR quantifies risk in terms of values of loss distribution [33]. Hence, it determines risk in terms of failure related to a decision to meet the predefined targeted profit. CDD is a percentile-based risk performance function. It maximizes total

Chapter - 2

expected rate of return of LSE's portfolio by including unequivocally portfolio drawdown into optimization [36]. DRC approach determines difference between target profit and scenarios with profit less than expected. This difference is called as downside risk [78]. Stochastic dominance risk management approach considers a set of profit profile that is dominant to pre-specified benchmark. Then, optimal profile that gives maximum expected profit is selected [79]. Here, benchmark is an estimation of profitability under uncertainty.

The measured risk can be controlled by selecting a particular trading strategy suitable to LSE. The chosen strategy would help to achieve the desired balance between risk and profit. The process of determining trade-off between risk and profit is called risk management [6]. Risk management approaches are diversification, hedging, and portfolio optimization [27]. Diversification is the process of allocating investment in various assets. In the context of portfolio, diversification indicates widening of portfolio, which attempts to maximize expected return for a given level of risk [80], [81]. Risk free assets such as bilateral contracts, forward contracts, and self-generation are preferred over spot market to minimize risk of profit variability [4], [82]. ESS and Demand Response Program (DRP) are new addition to risk free assets used for risk minimization.

Hedging uses derivatives, i.e., financial contracts to protect a company from a risky situation [45]. Availability of financial contracts is limited in the electricity markets, which restricts use of hedging. The 'best' approach to manage risk may differ from LSE to LSE. Risk management strategies to be adopted by LSE are influenced by its inherent risk-averse or risk-seeking nature [6]. Risk-averse LSE is more sensitive to risk than risk-seeking LSE. Therefore, it tries to minimize risk based on its sensitivity. However, risk-seeking LSE prefers to be exposed to a higher risk in the anticipation of higher expected profit. To avoid failures/ substantial financial losses or exploit benefits of risky situations, there is a need to prudently select suitable approaches for handling uncertainties and getting plausible and robust solutions.

2.6   Consumer Behavior Modeling Approaches

Consumer reacts to offered sale prices by modifying its demand. Such consumer behavior can be modeled by econometric, fuzzy, game theory, load profile, price elasticity, and bilevel programming approaches. These approaches are briefly described in the following subsections.

2.6.1 Econometric Approaches

Consumer behavior is modeled by a mathematical relation between demand and prices and can be represented by several functions. These functions may take the form of linear, power, or stepwise [8], [83]. The stepwise Price Quota Curve (PQC) is most common [18], [33], [56]. PQC is an econometric model, which captures and characterize consumers' willingness to pay, i.e., the quantum of electricity that a consumer is willing to buy at a certain price. Energy to be supplied by LSE decreases as its selling price increases [18]. Therefore, PQC function determines quantum of electricity to be supplied by LSE. However, PQC may be difficult to derive and does not explicitly model competition among rival LSEs [84], [85]. PQC assumes offered prices of rival LSEs and their risk attitude and strategies to be constant [46].

Market Share Function (MSF) is also based on an econometric model and analyses [24]. MSF shows the percentage of overall demand that can be served by LSE at different sale prices. MSF requires information about long-term strategy of LSE, risk attitude, consumer behavior, and rival LSEs behavior [43]. Moreover, information regarding the share of LSE and rival LSEs from loyal consumers, historical retail prices, consumers' search cost, switching cost, and allowed time period in which consumers may switch are required to model perfect competition [5], [46].

2.6.2 Fuzzy and Game based Approaches

Consumer switching tendency can be modeled by Fuzzy methodology [37]. Significant historical information on consumer switching, sale price information of rival and LSE under study is desirable to determine fuzzy membership function. However, it is difficult for LSE to identify if the consumer is loyal or a switcher [13]. Therefore, certain assumptions on consumer behavior are required. It is worth noting that PQC and MSF are intended to model consumer elastic behavior with respect to sale price so that, if the sale price to consumers is high, consumers would select a rival LSE [18], [43]. These functions do not provide opportunity to consumers to reduce their demand [47]. Alternatively, a game can be formulated among LSEs to model competition [84]. Such game can also be formulated as Nash equilibrium problem [46]. When Nash equilibrium

Chapter - 2

is achieved in competitive system, none of the LSEs have incentives to change their decision.

2.6.3 Load Profile based Approaches

An alternative way of modeling consumer behavior is to capture it from its load profile [32]. Based on load profiles, consumers are clustered into a number of classes using clustering methods such as k-means clustering. A typical load pattern is identified for these classes. Then acceptance function is defined to determine sale prices for each consumer class. This function is in some sense similar to PQC or MSF. Basically, acceptance function provides a trade-off between sale price and quantum of electricity that LSE can sale. Here, acceptance function is defined for a number of consumers and the maximum cost acceptable to them. Determining maximum acceptable cost is challenging as it depends on various parameters such as retail competition, economic situation, and weather conditions [86]. Moreover, clustering is subject to the method selected for clustering.

2.6.4 Price Elasticity

Price elasticity is another popular way to capture consumers' sale price responsiveness [23], [87]. This is because elasticity affects both electricity prices and scheduling of generation [47]. Elasticity measures the degree of response of demand to the changes in sale price. Hence, the ratio of the percentage variation of the quantity (energy demand) to the percentage variation of its (sale) price defines price elasticity of demand [87]. A price elasticity matrix consisting of self- and cross-elasticity describes consumer reactions. Self-elasticity relates demand during an hour to price during that hour. Cross-elasticity relates demand in one hour to the price during other hours [47]. Value of price elasticity signifies how much demand is elastic. It can be less or higher than one. Value of elasticity higher than 1 indicates that demand is elastic, whereas value of elasticity lower than 1 suggests that demand is inelastic. A negative sign in elasticity represents, price increase is accompanied by decrease in demand and vice-versa. Moreover, value of elasticity in context to electricity varies from one market to another, one consumer class to other consumer class, and one hour to other hours [88]–[90]. The variation in values of elasticity is due to the complexities of human behavior and geographical diversities [91]. Elasticity value is used to describe price sensitivity of consumers.

30

2.6.5 Bi-level Programing

In smart grid, consumers can flexibly and rationally control or schedule their demand depending on sale price to reduce electricity bill [92]. This is possible due to recent developments in energy management technologies. It enables LSE to communicate sale price signals to consumers well in advance to real-time electricity consumption [93]. According to received sale price signals and their operational constraints, consumers optimize their electricity consumption. Therefore, consumers have an economic goal of cost minimization, which is contradictory to LSE's goal of profit maximization [16]. This type of decision making is hierarchical in the sense that decision taken by LSE to optimize its goal is affected by the response of consumers, who seek to optimize its own outcomes. Such hierarchical decision making can be modeled as a Stackelberg relationship and is formulated using bi-level programming [94], [95]. PQC, MSF, and price elasticity methods cannot model hierarchical interactions between consumers and LSE.

In bi-level entirety, the two conflicting objectives, LSE's profit maximization, and consumers' energy bill minimization, become a joint optimization problem [16]. Consumers' optimization is expressed in the lower level of bi-level programming model. Suppressing/ shifting consumer's flexible demand modifies its welfare associated with comfort/ utility cost. Considering flexible demand and sale prices, consumers determine trade-off between energy bill minimization and welfare (comfort) maximization [96], [97]. However, consumers being more sensitive to (sale) price change seek to minimize energy bill irrespective of discomfort. Such LSE's decision making problems are considered for short or medium-term planning horizon [84], [97].

2.7 Demand Response and its Application

Fluctuation in consumer demand requires additional electricity procurement from spot market, which may be procured at high prices. This additional procurement increases cost of electricity. However, demand (load) is served at a flat rate. Flat rate is determined well in advance and indicates average cost of energy over the scheduling horizon [23]. This does not reflect any additional procurement made to supply actual demand. Hence, this leads to reduction of LSE's profit. Consumers have few or no opportunities to interact with electricity markets. Most consumers pay a flat price for electricity no matter when they use it, even during peak electric use when WEM prices skyrocket and

Chapter - 2

capacity shortages threaten system stability [18], [43], [84]. Weak communication infrastructure is a major hindrance to establish communication between consumers and LSE [93].

Technology advancement in the grid has helped to remove the barrier. This has provided opportunities to LSE to exploit demand (load) management program to avoid electricity procurement at a high cost. This program motivates consumers to reschedule their controllable demand to off-peak hours. In other words, consumers are encouraged to modify their consumption according to the prices in spot market. For this, some incentives are provided to consumers. This process of load management is called DR [98].

2.7.1 Demand Response Program

DRP has been employed in various forms to handle power system problems across the globe for decades [99]. DRP implementation is more active at the retail level than the wholesale level [100]. Economic motivation is a major driver of DRP. DRP is classified on the basis of either motivation and triggered criteria or the way of consumer enrollment to provide DR [101]. DRP classification includes Incentive Based DRP (IBDRP) and Price Based DRP (PBDRP).

2.7.1.1 Incentive Based Demand Response Program

IBDRP is also known as DR method with dispatch capability [102]. This is because consumers enroll themselves to participate in IBDRP. This program includes interruptible/ curtailable load contracts, Direct Load Control (DLC), Emergency DRP (EDRP), ancillary services, and demand bidding program [103]. Interruptible load and DRC are quite popular DRP utilized by LSE to manage consumers' demand. Under interruptible load contracts, consumer is asked to interrupt partial or full load during peak or critical hours [56]. Under DRC, utility company directly control heating and air conditioning load of consumers [101].

2.7.1.2 Price Based Demand Response Program

PBDRP is known as DR method without dispatch capability as consumers voluntarily participate in this program [102]. For PBDRP, dynamic sale prices are offered to consumers. This concept of DR through price signal was proposed in 1988 by Schweppe

[104]. It is quite popular in the current power systems to elicit consumer response through price signals. Dynamic sale prices are set to manage consumer demand profile. Dynamic pricing is briefly explained in a later section of this chapter. Low and high prices are offered during off-peak and peak hours to increase and decrease demand, respectively. Moreover, for price-based DRP, consumers do not enroll directly but react to sale price signals based on their price sensitivity or flexible demand. Benefits and challenges of DR in power systems from different entities' perspectives are extensively discussed in the literature [99], [101].

2.7.2 Demand Response Contracts

DR contracts are regarded as negative generation, to meet supply [99]. DR can be procured from DR aggregators besides consumers. DR contracts such as pool-order option, spike order option, forward-DR, and reward-based DR can be considered by LSE [105]–[107]. Pool-order option and spike order option are financial contracts that provide rights but not an obligation to exercise or buy them from the seller. Forward-DR and reward-based DR are agreements signed at fixed price or reward for fixed quantity of DR. As DR provider aggregates DR, it is responsible for implementing the DR program. Hence, LSE is not involved in DR aggregation. It only strategically utilizes these contracts to determine profit maximization decisions. DR contracts reduce procurement cost and price uncertainty to certain level, but might significantly affect revenue of LSE.

2.7.3 Application of Demand Response

LSE can exploit DR to achieve a specific goal to maximize its profit. This goal can be to reduce procurement from costliest forward contract [23], [24], reduce risk exposure of spot market price volatility [108], [109], or best utilization of RE sources [101]. LSE can effectively utilize DRP considering distribution network operation efficiency, to manage peak load [110]. In other words, tailor-made dynamic prices are required to achieve targeted goals of LSE.

2.8   Impact of Retail Competition on Consumer Switching

Transition of retail electricity supply from a monopoly operation to free competitive markets is a mandatory step from economic and social perspectives [10]. The retail

Chapter - 2

transition aims efficient operation of utility companies and delivers significant benefits to consumers. The achievement level depends upon retail market development. However, retail development varies largely between countries. Some countries are actively resisting change in retail structure while others are developing faster than required [11]. Development of retail markets has been very different in countries and hence the result of introducing competition into retail. Few countries have proven to be rather competitive and dynamic in several studies, whereas few countries have remained more neutral, at least when measured using the common competition indicators. Consumers switching rate is an important indicator for measurement of the level of retail competition. Switching rates are high in some countries, whereas it is very low in some. For example, Danish households exhibit low switching rate despite availability of cheap selling offers. This is because **Danish consumers' electricity bill constitute a** small proportion of their monthly income [111].

In a perfectly competitive retail market, sale prices are not determined and regulated or capped by regulatory body/ government. The end of regulated price has made it advantageous to purchase electricity from existing suppliers at rates fixed below market levels. Studies on impacts of deregulation on technical efficiency and changes for large power utilities in US indicates that deregulation has had a significant impact on generation efficiency, but not on distribution efficiency. However, for some countries which have undergone deregulation, consumer mobility has remained low. It is understood that LSEs are not interested in competition, and conditions for new entrants are very difficult. Retail prices are still regulated by regulatory body or government to protect consumers from varying prices. However, the range of products is rather wide in these countries. This necessitates investigations of LSE's decision making considering current market scenario.

2.9 Retail Pricing Schemes

LSE sells electricity to consumers at predetermined retail prices. Retail pricing is broadly classified into fixed and dynamic prices [24], [38]. Under fixed pricing, price remains fixed throughout the day and for at least several months [14]. Dynamic prices vary over time. These pricing are briefly described in the following subsections.

2.9.1 Fixed Pricing

Fixed or flat rate pricing is the most common retail pricing practiced all over the world before and even after deregulation [112]. This fixed price can be calculated based on the average of varying electricity procurement costs plus some profit margin and risk premium [101]. However, fixed sale prices do not reflect actual supply cost. Consumers are not exposed to the changing cost of electricity with a change in aggregated demand. Therefore, consumers do not face any risk of high electricity bills for any unplanned electricity consumption. Whereas, LSE has to procure, this unplanned quantum of electricity at varying prices, which would increase its procurement cost. Fixed pricing is considered as a limitation for LSE to exploit benefits of liberalized electricity markets. Inefficiency and limitation of fixed pricing can be reduced by dynamic pricing.

2.9.2 Dynamic Pricing

Dynamic price/ tariff structures or time varying prices have the potential to flatten/ modify consumer's demand profiles. This also provides an opportunity for consumers to reduce their electricity bills without changing consumption during the day, but just by shifting load from one hour to others. Under dynamic pricing, prices are varying over the period, and consumers are well informed about variations in prices [19]. However, dynamic pricing exposes consumers to risk resulting from WEM price fluctuation. The extent of risk exposure depends on design of pricing scheme. Dynamic pricing schemes can be designed based on various factors such as risk, rewards, and time (peak, off-peak, etc.) [19]. Moreover, dynamic pricing also depends on market structures [113]. Dynamic pricing can be classified into Time of Use (ToU), peak, and Real Time Pricing (RTP).

2.9.2.1 Time of Use Prices

ToU is a popular variant of time varying pricing, where electricity prices depend on the time at which electricity is consumed [14]. ToU prices should be based on principle of marginal cost pricing of electricity [112]. Hence, offered prices would be high during peak hours and low during off-peak hours to reflect the marginal cost of electricity. This pricing scheme is intended to reduce consumer's demand during peak hours and increase demand during off-peak hours [113]. Therefore, ToU reduces exposure of LSE

Chapter - 2

to price risk during peak hours of WEM, and consumers have an opportunity to save by redistributing their demand.

2.9.2.2 Peak Prices

Critical Peak Pricing (CPP), Variable Peak Pricing (VPP), and Peak Time Rebates (PTR) are variants of peak pricing [15]. These pricings are considered to reduce peak demand. CPP sets a higher energy price during hours of peak demand [39]. CPP is designed to reduce demand at critical peak hours [99]. Under this scheme, high rates are applied during critical peak demand events. Usually, CPP prices are higher than the peak of ToU scheme. CPP prices are informed to consumers hours or a day ahead before critical peak hours. An extension of CPP is VPP, which may consider daily variations of both peak hours and the price [19]. PTR pricing offers rebate for consuming below a certain predetermined level during peak hours.

2.9.2.3 Real Time Pricing

RTP closely reflects WEM prices. It accommodates fluctuation of WEM caused by weather conditions, generator failures, generation scarcity, or other contingencies that occur in WEM [112]. RTP completely transfers price risk exposure that arises due to fluctuation of WEM prices to consumers [20]. Various RTP models are designed from energy service providers [114], [115] and consumer's perspective to maximize their utility function [93]. RTP prices are set for day ahead or real-time. Day ahead RTP for all 24-hourly prices for a given day are announced at one time on the prior day. In real-time RTP, prices are announced on rolling basis, typically 15 to 90 minutes prior to the beginning of consumption hour [14].

Dynamic pricing scheme designs are governed by market structure [20], [113] and are tailored to meet a specific goal of the utility. For example, LSE with responsibility of network operation intends to reduce congestion, maintain smooth system operation, and maximize profit by offering dynamic pricing. However, LSE without network operation intends to offer dynamic pricing to reduce procurement cost. Another goal could be to effectively utilize high RE without curtailing it. This requires effectively designed tariff targeting high RE. Zero price ToU is suggested to achieve significant response from consumers to utilize high RE [116]. Prices can be designed from capacity investment or capacity cost perspective of LSE [117], [118]. A SO designs dynamic tariff to maximize welfare and improve system load factor [49]. Design of dynamic pricing requires knowledge about demand, demand-price relationship, and consumer behavior [20], [119]. These are necessary information for a supplier to plan its supply and tariff structures.

2.9.3 Risk Reward Mapping of Retail Pricing

The relative-risk reward situations between the various pricing scheme vary [15], [20]. These are developed to offer various choices to consumers considering their point of view. Flat rate pricing provides no risk and no reward. RTP could provide maximum risk and maximum reward to consumers when compared to the other types of time varying pricings (Fig. 2.4). Moreover, dynamic prices for each consumer can be customized considering its smart meter data [120]. This implies that number of pricing schemes could reach as high as the number of consumers. This may lead to serious price discrimination, which is too hard to implement in reality. Such pricing may decrease social welfare and lead to consumer exploitation.

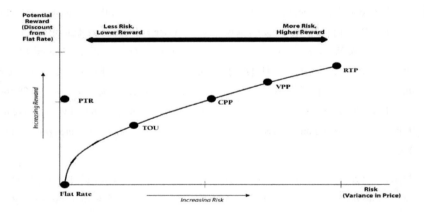

Fig. 2.4: Risk reward mapping of retail prices [15]

2.10 Impact of Renewable Energy Integration

The current trend around the world is to progressively integrate high RE into power systems. This RE integration aims to decarbonize electricity and reduce reliance on fossil fuels for energy supply. Various policies are being introduced in this regard world-wide. Mandatory targets of certain percentage of final energy consumption from

Chapter - 2

RE are set. To accomplish this, tax credit, emission allowance incentives, and feed-in tariffs are introduced. Some obligations are posed to suppliers such as Renewable Portfolio Obligation (RPO)/ Renewable Portfolio Standard (RPS) [121], [122]. RPO/ RPS places an obligation on suppliers to include some portion of power from RE generation in their energy procurement portfolio. RPO/ RPS is combined with the tradable green certificate. This allows suppliers to either procure RE or certificates. Green certificates are an alternative way to meet the RE obligation, without the physical procurement of RE.

RE integration has certain other advantages apart from decarbonization. RE connection at transmission or distribution level defers their expansion planning. As renewables typically rely on zero- or low-cost energy sources, efficiency of RE Sources (RES) is not a concern as compared to costly fossil-fuel based sources. However, capability and availability attributes of RES are of concern [123]. Prominent sources of RES such as solar and wind typically have pronounced seasonal and hourly profiles. Therefore, availability of these RES is intermittent. Non-dispatchable characteristics of solar and wind make it non-schedulable. However, use of thermal storage makes solar thermal dispatchable.

Despite their many advantages, renewables challenge the operation of electricity markets with their intermittent nature [124]. RE has impacted WEM prices as these units modify dispatch order of existing generation units [121]. WEM prices have become more volatile than ever due to the high penetration of intermittent RE [116]. RE also impacts network cost, system service cost, and taxes. Hence, retail price which includes these components are also influenced by RE [121]. Support costs for electricity from RES are recovered through retail prices. This has substantially increased retail prices [125]. Moreover, country-specific regulatory policy has a significant impact on retail prices. Energy imbalance management arising due to intermittent RE is another important challenge for market integration.

High RE in LSE's procurement portfolio supports ambitious government targets of decarbonization. Uncertain RE generation requires effective balancing of supply and demand. However, a mismatch between supply and demand would affect the procurement cost of suppliers [123] and consequently impact retail prices. RE uncertainty modeling can minimize its impact on decisions of the supplier. Also, aggregating RE from geographically spread plants is helpful to minimize RE variability. Flexible resources such as DR potential can be utilized to manage imbalances arising due to intermittent RE [107]. Non controllability and dispatchability of RE force LSE to curtail RE during the low demand period. This restricts LSE to potentially utilize available RE. For such conditions, DRP can be exploited to utilize RE to a certain extent. As it is mentioned before that retail prices are influenced by RE, potential research in designing effective tariff/ pricing considering various aspects of RE is highly desirable. As a whole, RE significantly impacts LSE's profit and risk management strategies. This necessitates development of adequate models to devise prudent decisions to maximize LSE's profit.

2.11 Research Challenges

LSE's operations in the liberalized electricity market face several challenges. Decision making on procurement and sale prices under market uncertainties, consumer behavior, and renewable are considered in this thesis. Significant research on LSE's decision making has been conducted in recent years. These indicate that dynamic sale price determination targeting risk-based decision making is a potentially challenging problem. On the basis of critical literature review pertaining to LSE's decision making, identified research challenges have been shown in Fig 2.5 and stated as follows.

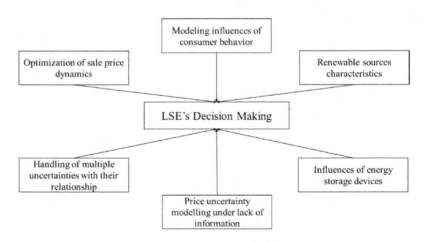

Fig. 2.5: Identified research challenges

Chapter - 2

- Optimization of sale price dynamics for LSE under spot market price uncertainty, consumer elastic demand, and renewable energy.
- Consumer's price responsive behavior modeling through bi-level programming for LSE's profit maximization decision making.
- Spot market price uncertainty modeling under lack of information to maximize LSE's profit for risk and opportunity situation.
- Determination of LSE's profit maximization strategy considering energy storage and renewables.
- Modeling of correlation between spot market price and demand to maximize LSE's profit and manage spot market price risk.

Chapter – 3: LSE's Risk Based Profit Maximization

This chapter presents a decision making model to determine sale prices and procurement decision for elastic consumer demand aiming to maximize LSE's profit.

Chapter - 3
LSE's Risk Based Profit Maximization

3.1 Introduction

LSE's profit depends on its procurement cost from WEM and revenue from sale of electricity to consumers. As discussed in Chapter 2, for price responsive consumers, procurement and sale price decisions of LSE are affected by consumer behavior, uncertainties associated with wholesale and retail markets, and structure of retail market. In a fully competitive retail market, LSEs try to maintain or increase their market share by strategic decision making [10], [126]. However, the market share of incumbent LSE is minimally impacted due to low consumer switching rate. Incumbent LSE offers attractive dynamic pricing to exploit opportunities of demand shifting [19]. This necessitates the modeling of elastic demand of consumers to determine optimal dynamic sale prices. On the procurement side, an important concern for LSE is spot market price variability [6]. Such concern can be managed by offering cost-reflective dynamic sale prices to consumers. This necessitates determination of dynamic sale prices targeting consumers' elastic demand and spot market price variability.

Consumer response to sale price modifies its demand profile. As a consequence, demand to be procured by LSE at any instant is modified [9]. Therefore, change in demand over different hours significantly impacts LSE's energy procurement cost. This needs to be considered while sale price is determined.

Moreover, variability and availability of RE sources such as solar and wind would affect energy procurement and sale price decisions of LSE. LSE without RE self-generation generally focuses on spot market price and consumer behavior to maximize its profit [18], [43]. When RE self-generation, particularly solar, is included in procurement portfolio, LSE may target to best utilize energy from RE based on its availability rather than from spot market, to maximize its profit. LSE's priority for RE utilization may affect strategy for procurement and sale of electricity. Hence, investigations on LSE's decision making considering RE availability are relevant.

As spot market prices exhibit extreme volatility than other uncertain parameters [17], [127], its impact, i.e., risk needs to be minimized while LSE makes procurement and

Chapter - 3

dynamic sale price decisions. Therefore, risk-constrained optimization models are required to maximize profit and minimize spot market price risk [1].

This chapter presents a benchmark model to determine optimal dynamic sale prices and energy procurement portfolio for an LSE, with and without RE utilization. Consumer's elastic demand and spot market price uncertainty are modeled to maximize LSE's profit. Risk imposed by spot market prices in energy procurement portfolio is minimized through a trade-off between profit and risk. This risk is modeled and assessed using CVaR. Change of consumer demand in response to offered sale prices is modeled using demand function. To illustrate the impact of consumer behavior on LSE's decision making, three major consumer classes such as residential, agricultural, and industrial are considered. Further, impact of high solar energy and its availability on dynamic sale prices is explicitly investigated in this chapter.

3.2 Problem Description

This work considers medium-term decision making of LSE to maximize its profit. LSE maximizes profit by maximizing revenue from selling energy at offered dynamic sale prices and minimizing total procurement cost from various available resources. Considering incumbent LSE as large, its overall market share is assumed not to vary, despite temporal demand shift, i.e., overall energy consumption of consumers over the horizon remains constant. However, its consumer demand is elastic and responsive to offered sale prices [88]. Spot market price being extremely volatile than demand [17], [36], this work emphasizes on considering price uncertainty to analyze its influence on LSE's expected profit. Considering elastic behavior of consumer demand and spot market price uncertainty, LSE makes decisions for i) Energy procurement portfolio (spot market, bilateral contracts, and self-generation) ii) Optimum value of dynamic sale prices to achieve higher profits. The decision making problem is shown in Fig. 3.1.

LSE determines dynamic sale prices and informs these pricing signals to consumers. Consumer demand would respond to offered sale prices, as per its elasticity. Optimal energy procurement portfolio for demand aims to minimize procurement cost by the best utilizing available options. Hence, both demand (effective) and sale price to be offered are to be decided to maximize LSE's profit (Fig. 3.1). However, as the decision making is done well in advance, energy price from spot market cannot be known deterministically at the time of decision making, leading to risk of spot market price volatility. Hence, the risk of spot market price needs to be considered for optimal decision making [6]. Profit maximization and risk minimization have been modeled using CVaR to obtain an optimal trade-off between profit and risk.

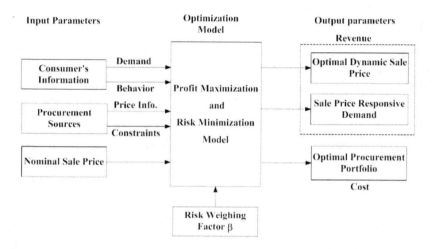

Fig. 3.1: LSE's decision making

This work considers 24 time blocks to precisely highlight the impact of various input parameters such as demand and wholesale prices, on each hour. LSE provides electricity to different consumer classes; residential, agricultural, and industrial. Consumers, as a part of different classes, have different consumption patterns and elastic nature [88]. Detailed knowledge of consumer's behavior can facilitate LSE to determine specific tariff to maximize its profit. Hence, dynamic sale prices for different consumer classes are evaluated and compared.

3.3 Mathematical Model

For the given description in Section 3.2, LSE's problem is formulated in two segments, i.e., profit maximization and risk minimization. LSE profit is modeled as the difference between generated revenue and procurement cost. Risk is modeled using CVaR.

3.3.1 Procurement Cost

LSE's procurement cost is the sum of cost for energy procured from bilateral contracts, thermal and RE self-generating units, and spot market in hour t. Total procurement cost is given as

Chapter - 3

$$Cost_t^{total} = C_t^B + C_t^{SG} + C_t^S \tag{3.1}$$

3.3.1.1 Bilateral Contracts

LSE signs bilateral contracts with generating companies at a mutually agreed fixed price $\lambda_{i,t}^B$ for a specified time t. Let total number of available bilateral contracts be I. Cost of energy procured from bilateral contracts C_t^B in hour t is given by (3.2). Minimum and maximum procurement limits on bilateral contracts are given by constraint (3.3).

$$C_t^B = \sum_{i \in I} P_{i,t}^B \lambda_{i,t}^B \tag{3.2}$$

$$P_i^{min} \delta_{i,t} \le P_{i,t}^B \le P_i^{max} \delta_{i,t} \tag{3.3}$$

3.3.1.2 Self-generation

LSE uses two types of self-generation as procurement options. One is conventional thermal generation, and another is renewable based solar Photovoltaic (PV) generation. At time t, total energy procured from self-generation would be equal to energy procured from thermal and PV generation.

$$P_t^{SG} = \sum_g P_{g,t} + PV_t^{act} \tag{3.4}$$

3.3.1.2.1 Thermal Generation

Operational cost of thermal generation C_t^{TG} is considered based on fuel cost, start-up, and shut down costs. Cost from thermal self-generating units is

$$C_t^{TG} = \sum_g \left(Cost_{g,t}^F + c_{g,t}^{su} + c_{g,t}^{sd} \right) \qquad \forall g \tag{3.5}$$

Fuel cost $Cost_{g,t}^F$ to procure energy from g^{th} thermal generating unit in hour t is given by (3.6), subject to constraints (3.7) to (3.15). Quadratic cost function is linearized by piece-wise linearization as given in Appendix-A.

$$Cost_{g,t}^F = a_g \left(P_g^{min} \right)^2 + b_g P_g^{min} + c_g u_{g,t} + \sum_k s_g^k P_{g,t}^k \tag{3.6}$$

$$c_{g,t}^{su} \ge C_g^{su} (u_{g,t} - u_{g,t-1}) \qquad \forall g, \forall t \tag{3.7}$$

46

$$c_{g,t}^{sd} \geq C_g^{sd}(u_{g,t-1} - u_{g,t}) \qquad \forall g, \forall t \qquad (3.8)$$

$$P_{g,t} - P_{g,t-1} \leq R_g^u . u_{g,t} \qquad \forall g, \forall t \qquad (3.9)$$

$$P_{g,t-1} - P_{g,t} \leq R_g^d . u_{g,t-1} \qquad \forall g, \forall t \qquad (3.10)$$

$$[X_{g,t-1}^{ON} - T_g^U].[u_{g,t} - u_{g,t-1}] \geq 0 \quad \forall g, \forall t \qquad (3.11)$$

$$[X_{g,t-1}^{OFF} - T_g^D].[u_{g,t-1} - u_{g,t}] \geq 0 \quad \forall g, \forall t \qquad (3.12)$$

$$P_g^{min} . u_{g,t} \leq P_{g,t} \leq P_g^{max} . u_{g,t} \qquad \forall g, \forall t \qquad (3.13)$$

$$c_{g,t}^{su}, c_{g,t}^{sd} \geq 0 \qquad \forall g, \forall t \qquad (3.14)$$

$$u_{g,t} \in [0,1] \qquad \forall g, \forall t \qquad (3.15)$$

Constraints (3.7) and (3.8) decide start up and shut down cost of self-generating units. Constraints (3.9) and (3.10) decide ramp up and ramp down limits. Minimum up and down time of generating units is given by constraints (3.11) and (3.12), respectively. Minimum and maximum generation limits are decided by constraint (3.13). (3.14) is a non-negativity constraint. (3.15) is variable declaration constraint.

3.3.1.2.2 Renewable Generation

Marginal cost of energy generation from RE based PV systems is negligible. Therefore, LSE neglects cost of energy procurement from PV systems while decisions on sale price and energy procurement are made. However, LSE subtracts energy procurement cost of PV generation from expected profit to show cost of investment and maintenance. The cost of energy procured from PV systems is considered using (3.16). This equation was not utilized for optimization problem. The value of λ_t^{PV} can be mathematically calculated using bottom-up or any other approach, by considering relevant cost parameters of PV systems. Calculation of λ_t^{PV} is beyond the scope of the work, so its value is assumed. Also, PV generation is curtailed during periods of low demand considering (3.17).

$$C_t^{PV} = PV_t^{act} \lambda_t^{PV} \qquad (3.16)$$

$$PV_t^{act} = PV_t - PV_t^{cur} \qquad (3.17)$$

Chapter - 3

3.3.1.3 Spot Market

LSE considers spot market as one of the procurement options. Future prices are uncertain in spot market. Therefore, forecasted/ expected values of spot market prices are considered. Expected energy procurement cost from spot market for each period t is given by (3.18). Constraint (3.19) restricts energy sale in spot market, i.e., only energy procurement from spot market is possible. This consideration is made, as medium-term procurement decisions are taken prior to actual delivery of energy. Also, this helps to precisely analyze the impact of consumers' behavior on LSE's decision making.

$$C_t^S = P_t^S \lambda_t^S \tag{3.18}$$

$$P_t^S \geq 0 \tag{3.19}$$

3.3.2 Energy Balance Constraint

Energy balance constraint (3.20) states that energy procured from various sources should match consumer demand at each hour t. Here consumer demand is consumer's sale price responsive demand L_t^{act}, as obtained from (3.22).

$$P_t^S + \sum_{i \in I} P_{i,t}^B + P_t^{SG} = L_t^{act} \tag{3.20}$$

3.3.3 Revenue

LSE generates revenue by selling energy to its group of consumers at offered sale prices λ_t^{sale}. Revenue for each time slot is

$$Revenue_t = L_t^{act} \lambda_t^{sale} \tag{3.21}$$

Both demand and sale price are decision variables of the problem and calculated by solving proposed optimization problem. Here, revenue depends upon sale price and revised demand. However, the allowable reduction in demand from its original forecasted value L_t is as per price elasticity ε_t.

3.3.4 Sale Price Modeling

Relation between consumer demand and sale price is modeled by linear demand function [8]. This model is the simplest and most widely used model [9], [24]. This function (3.22) shows a relation between sale prices λ_t^{sale}, modified demand L_t^{act}, price elasticity ε_t, nominal sale prices λ_t^{nsale}, and forecasted demand L_t. Hourly self-elasticity describes consumers' response for change in demand to change in price in that hour. Demand change is negatively related to price change. Thus, price increase leads to demand decrease and vice-versa. As focus of dynamic sale prices is to encourage consumers to shift their demand to periods (hour) of low spot market hours, consumers' predefined price responsiveness behavior between hours is not considered. Therefore, overall change in demand is subject to constraint that consumers' total demand throughout a day remains same. This is imposed by constraint (3.23). Thus, demand function determines equilibrium between sale price and demand that maximizes LSE's profit.

$$L_t^{act} = L_t(1 + \varepsilon_t \frac{\lambda_t^{sale} - \lambda_t^{nsale}}{\lambda_t^{nsale}}) \tag{3.22}$$

$$\sum_t L_t^{act} = \sum_t L_t \tag{3.23}$$

Nominal value of dynamic sale prices is evaluated considering flat energy price and spot market prices. Constraint (3.24) restricts LSE's sale prices from going below a certain percentage of nominal sale price. ψ_t^{min} in (3.24) is the value in percentage of nominal sale price λ_t^{nsale}, decided by LSE. Maximum sale price λ_t^{max} is regulatory price cap set according to market norms (3.25). For considered planning horizon, consumer switching is rare and thus neglected.

$$\lambda_t^{sale} \geq (1 - \psi_t^{min}).\lambda_t^{nsale} \tag{3.24}$$

$$\lambda_t^{sale} \leq \lambda_t^{max} \tag{3.25}$$

3.3.5 Price Uncertainty Modeling

Spot market prices are uncertain and can not be predicted with certainty beforehand at the time of decision making. Uncertainty in spot market prices introduces risk for LSE. This risk in LSE's energy procurement is quantified and measured using CVaR. Risk measure CVaR is defined as the conditional expectation of losses that exceeds a threshold value (VaR) (Fig. 3.2) [6]. In other words, for a confidence level α', CVaR is defined as the expected profit of the $(1-\alpha')*100\%$ scenario with low profit. VaR provides expected maximum loss over a target horizon within a confidence interval. CVaR at a given confidence level is the expected loss given that the loss is greater than VaR at that level. CVaR is computed as the weighted average of losses exceeding VaR values [28] (Fig. 3.2).

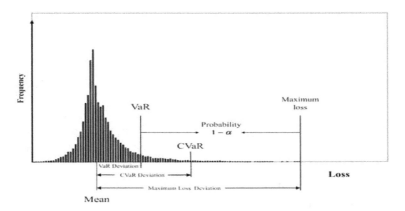

Fig. 3.2: Conditional Value at Risk

Risk incorporated by spot market procurement is accounted by (3.26). $CVaR(\lambda_t^S)$ represents calculated CVaR for per unit procurement.

$$Risk_t = P_t^S . CVaR(\lambda_t^S) \qquad (3.26)$$

In this work, historical simulation method of CVaR computation is applied [27]. CVaR computation based on historical simulation method is modeled using following algorithm [27] and obtained value of $CVaR(\lambda_t^S)$ is substituted in 3.26.

1. Compose an array of historical spot market prices for considered (historical) period.
2. Calculate an array of the changes in spot market prices (value) from the previous day's prices.
3. Arrange price values in sequence from largest positive to the largest negative.
4. Take an average of all values exceeding threshold value (VaR).

3.3.6 Objective Function

Overall objective for the LSE is to maximize expected profit and minimize involved risk. This multi-objective optimization problem calculates trade-off between expected profit and involved risk, by combining the two objectives with a risk weighting factor β. β increases from 0 to a higher value with risk-averse nature of LSE. $\beta = 0$ reflects risk-neutral behavior, i.e., LSE concern for profit maximization only, without any risk consideration. LSE's risk-averse nature is reflected by values of β. The overall objective function can be written as

$$Maximize \quad Obj = \sum_t Prof_t - \beta \sum_t Risk_t \tag{3.27}$$

In (3.27), $Prof_t$ is LSE's expected profit, and is calculated by subtracting procurement cost (3.1) from generated revenue (3.21) as shown in (3.28). $Risk_t$ signifies risk (3.26), which is to be minimized, and is thus indicated with a negative sign.

$$Prof_t = Revenue_t - Cost_t^{total} \tag{3.28}$$

Objective (3.27) is maximized subject to constraints on bilateral contracts (3.3), self-generating (3.4)-(3.17), spot market (3.19), energy balance (3.20), sale price and demand redistribution constraint (3.22)-(3.25). Here, sale price, price-responsive demand, and portfolio of energy to be procured from various options are key decision variables of the optimization problem.

3.4 Case Study

3.4.1 Data

Proposed model is illustrated *via* a case study for an assumed large incumbent LSE, using historical spot market price data of December 2016 - March 2017 from Pennsylvania-New Jersey-Maryland (PJM) electricity market [128]. As large LSE's market share is fixed, retail market consideration is not relevant. LSE decides dynamic sale price and WEM procurement for a planning period of 1 day, considering an hour as a time block for sale price. One bilateral contract at 34$/MWh is considered with maximum and minimum trading limits as 500 MW and 30 MW respectively. Self-generation consists of thermal and PV based RE-generation. LSE owns single thermal generating unit whose specification is given in Table 3.1. PV generation profile is obtained for the Pennsylvania region [129]. To generate this profile, other system operating parameters such as efficiency, tilt angle, and azimuth are considered equal to

Chapter - 3

90%, 35°, and 180°, respectively [129]. Procurement from PV generation is carried out at 27 $/MWh.

Table 3.1: Specifications of thermal self-generation unit

Quantity	Value	Unit
Capacity	130.00	MW
Minimum Power Output	20.00	MW
Ramping Limit Up	80.00	MW/h
Ramping Limit Down	80.00	MW/h
Quadratic Cost	0.03	$/(MW)^{2}h
Linear Cost	28.00	$/MWh
No-Load Cost	400	$/h
Start Up Cost	200	$
Shut Down Cost	100	$
Minimum Up and Down Time	1	h

To independently highlight the impact of RE generation on dynamic sale prices and spot market prices, two cases have been considered as

(i)   Case-A: Proposed model without RE generation and

(ii) Case-B: Proposed model with RE generation.

This work considers three consumer classes: residential, agricultural, and industrial, to understand their behavior on LSE's procurement and sale price strategies. This classification can be determined by grouping consumers with similar energy consumption patterns. LSE sells energy to each of these consumer classes, for their forecasted demand, at sale price for 24 blocks. Each consumer class has specific responsiveness to offer prices, according to their price elasticity. Elasticity computation is beyond the scope of this work and its values are assumed from the literature [88]. Price elasticities for residential, agricultural, and industrial consumer classes are considered as -0.8, -0.5, and -0.4, respectively. Assumed demand profiles for consumer classes are given in Fig. 3.3. These demand profiles are assumed based on usual energy consumption patterns for each class. Based on typical regulatory rules on tariffs, flat price for each class is considered as an average of expected spot market prices. Calculated flat price for each consumer class is equal to 29.67 $/MWh. Nominal prices for each class are determined by considering prices 5% higher than average spot market prices. Percentage variation for each consumer class is considered the same. A price range cap of 17 to 54 $/MWh is assumed to keep sale price in a reasonable limit for all consumer classes.

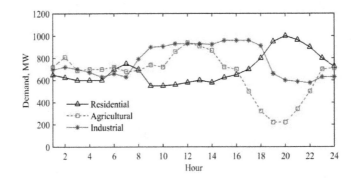

Fig. 3.3: Consumer classes' demand profile

3.4.2 Statistical Calculations

For each hour, expected values of spot market prices are calculated as average of historical prices for that hour, over entire planning period. Historical spot market prices are used to assess CVaR for per MW energy procurement from spot market for each hour. Fig. 3.4 depicts values of expected spot market prices and CVaR computed for 95% confidence interval over 24 hours.

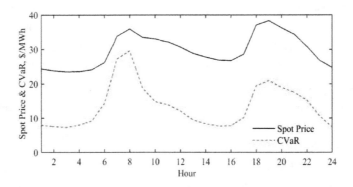

Fig. 3.4: Hourly spot market price and corresponding CVaR

Chapter - 3

3.4.3 Simulations

Simulations are performed to maximize (3.27), subject to constraints (3.3-3.17), (3.19), and (3.22-3.26), for different values of risk weighing factor β. Those are repeated for both cases and all consumer classes. The Mixed Integer Non-linear Programming (MINLP) problem has 409 (457) variables and 48 (48) binary variables for Case A (Case-B). It has been solved using industry grade software GAMS® using its solvers SBB CONOPT3 on Intel®, Core$^{(TM)}$, i5 CPU, 3.20 GHz, and 4GB of RAM system [130].

SBB, a known solver for MINLP solutions, works with NLP sub-solver, is based on a combination of the standard Branch and Bound (B&B) algorithm to provide MIP solution to NLP sub-solver [131]. CONOPT, an efficient solver for large NLP models, utilizes solutions obtained by SBB, to solve NLP sub-problem by a generalized reduced gradient algorithm [131]. Average execution and computation time to solve the optimization problem with three consumer classes is given in Table 3.2. It is observed that execution time increases with β.

Table 3.2: Execution and computational time

Consumer Class	Average Execution Time (s)		Average Computation Time (s)	
	Case-A	Case-B	Case-A	Case-B
Residential Consumer	0.113	0.269	2.796	3.437
Agricultural Consumer	0.069	0.061	2.704	3.266
Industrial Consumer	0.142	0.073	3.296	3.219

3.4.4 Results

Results obtained for three consumer classes: sale price, price-responsive demand, and purchase from different trading options are illustrated in subsequent sub-sections. Further, impact of demand profiles and price responsiveness of all consumer classes on dynamic sale prices and consumer demand is analyzed. The comparative assessment addresses the risk-neutral and risk-averse nature of LSE.

3.4.4.1 Case-A: Proposed Model without Renewable Generation

In this case, LSE utilizes thermal generation apart from WEM to obtain optimal procurement portfolio. Thus, in Eq. (3.4) PV_t^{act} would be zero. The impact of spot market prices and consumers' elastic demand on dynamic sale price is presented.

3.4.4.1.1 Impact on Dynamic Sale Price

As shown in Fig. 3.5, dynamic sale prices generally follow spot market prices. Due to high value of sale price, consumer demand tends to shift from high price periods to low price periods, according to elastic nature of each consumer class. Such observations can be made by analyzing Figs. 3.5 and 3.6 for each consumer class. This demand shift is primarily governed by consumer behavior and impacts LSE's profit.

Demand profiles (Fig. 3.3) and price elasticity for different consumer classes impact sale price calculations differently. For residential consumers, peaks and valleys of market prices coincide with their demand pattern. Thus the obtained sale price offers a demand shift as per elasticity of consumer class. The offered sale prices and revised demand is seen in Figs. 3.5a and 3.6a, respectively. For agriculture consumers, sale prices are high during off-peak (Hours 17 and 23) of demand profile, as spot market has high prices (Figs. 3.5b and 3.6b). During these hours, agriculture demand is low, still, high sale prices restrict consumers from increasing their demand. Also, for such cases, demand during Hours 17 and 23 does not shift to other hours even though high prices are offered, as demand is already at its minima, and as per Eq. (3.22), impact of price change on demand is minimum. For industrial consumers, similar observations can be made from Figs. 3.5c and 3.6c, as it also has a lower demand during the peak of spot market.

Due to high elasticity, consumption change for residential consumers is more as compared to other consumer classes. Maximum consumption decrement during 24 hours for residential, agricultural, and industrial consumer classes is 10.96%, 10.53%, and 12.59%, respectively. However, maximum increment in demand during 24 hours for residential, agricultural, and industrial consumer classes is 11.84%, 11.46%, and 9.88%, respectively. The revised demand curves are flattened, which helps LSE to reduce high price purchase.

Chapter - 3

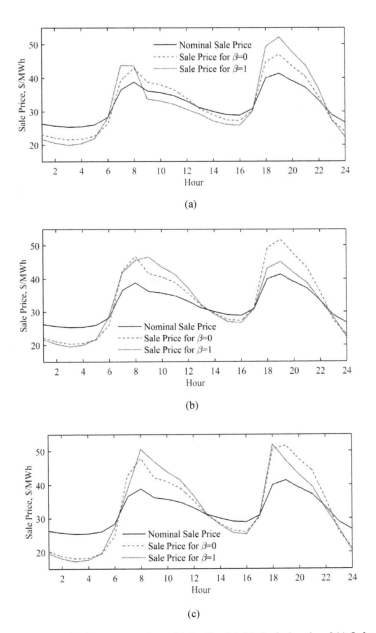

Fig. 3.5: LSE sale price for consumer classes (a). Residential, (b). Agricultural, and (c). Industrial

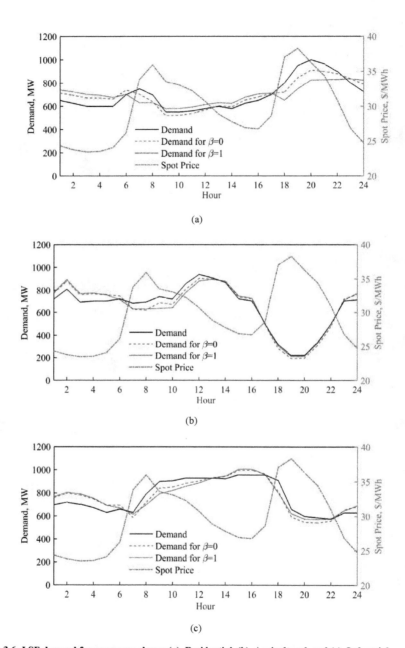

Fig. 3.6: LSE demand for consumer classes (a). Residential, (b). Agricultural, and (c). Industrial

Chapter - 3

Consumer's responsiveness to offered price leads to demand shifting from high price periods to low price periods. This reduces LSEs' total energy procurement cost and its spot market price risk. LSE's aim of risk minimization, as the second objective of (3.27), is discussed hence to assess the impact of risk-averse nature of LSE.

3.4.4.1.2 Impact of Risk Averseness on Decision Making

Sale prices and demand *w.r.t.* risk-neutral and risk-averse LSE for various consumer classes is represented by Figs. 3.5 and 3.6. As per (3.26), when LSE targets spot market price risk minimization, demand shifts during time blocks (hours) with relatively low volatility and low prices. However, sale prices tend to increase during time periods when spot market prices are highly volatile to address its risk-averse behavior. In Fig. 3.4, Hours 7, 8, and 19 with very high values of CVaR, have high sale prices. Consequently, demand goes low during those hours. Hence, high prices and demand situations can be avoided. With increasing risk aversion level (higher β), the impact of risk aversion increases. Though for low spot market price risk periods, impact of risk aversion is not so prominent. The maximum risk aversion level for the considered case is obtained at a value of β equal to 1. This value of β provides maximum possible optimal trade-off between profit and risk. Beyond this value, LSE incurs losses because procurement cost becomes higher than revenue. Therefore, for the considered case study, β equal to 1 could be a last viable choice for a risk-averse LSE.

For residential consumer class, spot market price profile coincides with their demand profiles to avoid most risky position of high market prices and high demand during Hours 17 to 23. Thus, sale price is high, resulting in large demand shifting. However, during these hours, agricultural and industrial consumer classes have little demand shift due to opposite demand profile. It is to be noted that for agricultural and industrial consumers, sale prices offered by risk-averse LSE are lower during their low demand period, as compared to a risk-neutral LSE. The reason being a low demand can be satisfied only from bilateral contracts avoiding procurement from high spot market prices. Therefore, sale prices follow bilateral contract prices. Conclusively, demand shifts to high demand hours, when spot market purchases cannot be avoided.

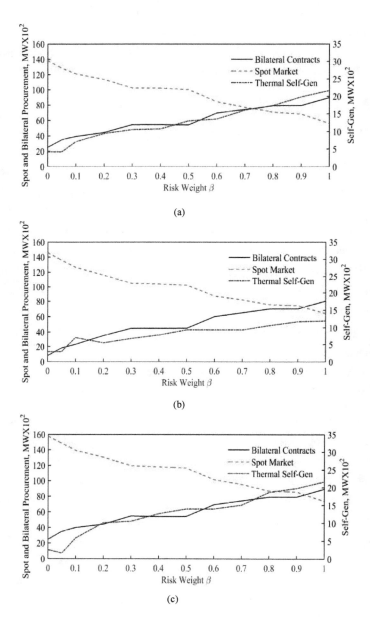

Fig. 3.7: Purchase from various procurement options for consumer classes (a). Residential, (b). Agricultural, and (c). Industrial

Chapter - 3

Fig. 3.7 shows optimal total purchase from various options (bilateral, self-generation, and spot market), for all consumer classes. With increasing β, results show that procurement shifts towards risk-free options to address risk-averse behavior, i.e., from the spot market to bilateral contracts and self-generation. However, this leads to increase in procurement cost (Fig. 3.8), as less risky options are relatively costlier and therefore, LSE's profit reduces (Fig. 3.9). As risk impacts both revenue and procurement decisions, variation between risk (CVaR) and expected profit for various consumer classes are demonstrated by efficient frontier (Fig. 3.9).

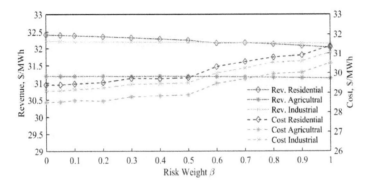

Fig. 3.8: LSE's revenue and cost for three consumer classes

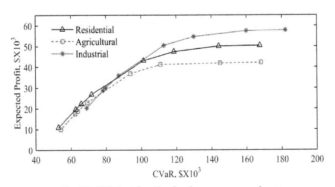

Fig. 3.9: Efficient frontier for three consumer classes

A comparison of expected profit with nominal and optimal dynamic sale prices is shown in Table 3.3, for three consumer types. It could be observed that optimized dynamic sale prices generate higher expected profit than nominal sale prices.

LSE's Risk Based Profit Maximization

Table 3.3: Expected profit under nominal and optimal dynamic sale price

Consumer Type	Profit under Nominal Sale Price ($)	Profit under Sale Price ($)	% Change in Profit
Residential	46623.74	50422.05	8.15
Agricultural	38281.75	42155.43	10.11
Commercial	50209.49	57868.10	15.25

3.4.4.2 Case-B: Proposed model with Renewable Generation

In this case, LSE utilizes thermal and PV generation, along with WEM, for optimal procurement portfolio. The results obtained for this case represent a significant impact on dynamic sale price.

3.4.4.2.1 Impact of Renewable on Risk-Averse Decision Making

Evaluated dynamic sale price and price-responsive demand for three consumer classes are shown in Figs. 3.10 and 3.11. Considering PV generation, sale prices are changed to shift demand to accommodate generation mix. Sale prices are reduced during periods of renewables availability. This reduction in sale price incentivizes consumer to shift significant quantum of demand from other hours to renewable availability hours. Significant shift in peak demand to renewable procurement hours for residential consumer can be clearly observed in Fig. 3.11a. However, a dip in demand at Hour 8 and 18 is observed due to high sale price. For agricultural consumer class, escalation in demand peak is observed during its demand profile peak (Fig. 3.11b). This happens because of renewable procurement, resulting in low sale price. On the contrary, sale price increases during off-peak period of demand profile. This helps to compensate for a significant decrease in demand. A similar observation for industrial consumer can be noticed in Figs. 3.10c and 3.11c. For both consumer classes, its peak demand coincides with renewable procurement peak. Further, Fig. 3.11 shows that considered PV generation is curtailed to the level of consumer demand for three consumer classes. However, PV curtailment is done, as consumers cannot shift their entire demand to renewable availability hours.

Chapter - 3

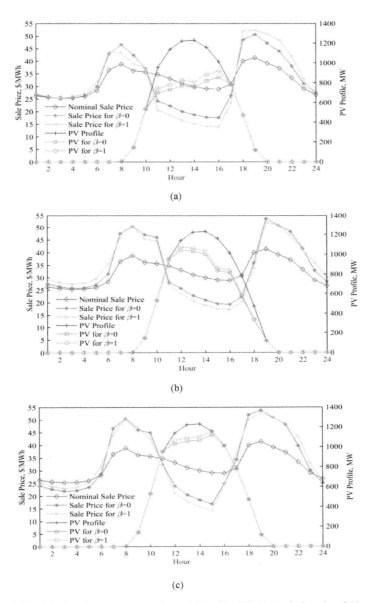

Fig. 3.10: LSE sale price for consumer classes (a). Residential, (b). Agricultural, and (c). Industrial

LSE's Risk Based Profit Maximization

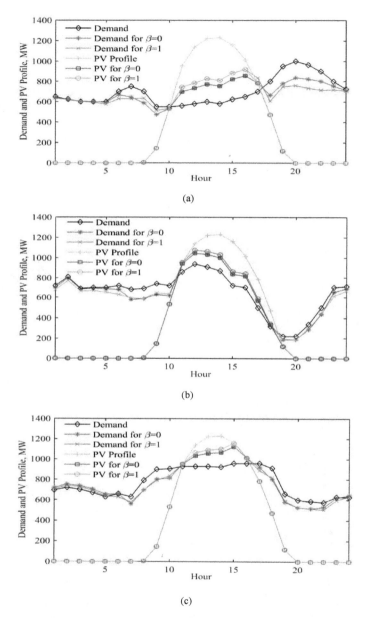

Fig. 3.11: LSE demand for consumer classes (a). Residential, (b). Agricultural, and (c). Industrial

Chapter - 3

Figs. 3.10 and 3.11 also highlight impact of renewable sources on risk-averse decision making of LSE. Figs. 3.10 and 3.11, illustrate that risk-averse (high value of β) LSE procures more from less risky sources, i.e., bilateral contracts and self-generation. For high values of β, evaluated risk and sale prices are low during renewable availability hours and high during non-renewable availability hours. This increases procurement from RE and reduces from spot market, thus reducing spot market price risk. For agricultural and industrial consumers, risk-averse LSE offers sale prices similar to risk-neutral LSE during off-peak hours of its demand profile. This happens because agricultural and industrial consumers' demand profiles do not follow spot market profile and have low elasticity as compared to residential consumers.

Fig. 3.12 shows the optimal purchase from various options (bilateral, self-generation including thermal and PV, and spot market) for three consumer classes. With increasing β, energy procurement increases from bilateral and self-generation to address risk-averse behavior of LSE. This would increase procurement costs. As compared to Case-A, LSE significantly reduces procurement from other options to meet demand.

Evaluated results highlight that dynamic sale prices are selected based on LSE's exposure to high spot market prices, and risk-averse behavior to avoid risk. Offered dynamic sale prices reduce peak demand requirements during high price periods, as elastic consumer's demand is responsive to offered prices. Redistribution of consumer demand reduces risk and costly energy procurement during peak hours. Therefore, dynamic sale prices contribute to procurement cost minimization and reduction in associated risk for LSE, due to demand shifting from high prices to lower price hours. Further, LSE's strategy with renewable procurement shows that renewable procurement prominently modifies LSE's procurement and selling strategies and increases its profit. LSE significantly reduces sale prices during renewable procurement hours and a consequent shift in peak demand to renewable availability periods is observed.

Fig. 3.13 highlights efficient frontier for three consumer classes. Expected profit and corresponding risk decrease with increase in value of β. For high values of β, LSE increases procurement from risk free options to reduce risk. These options are costlier than spot market. Hence, increases procurement cost and thereby reduces profit.

LSE's Risk Based Profit Maximization

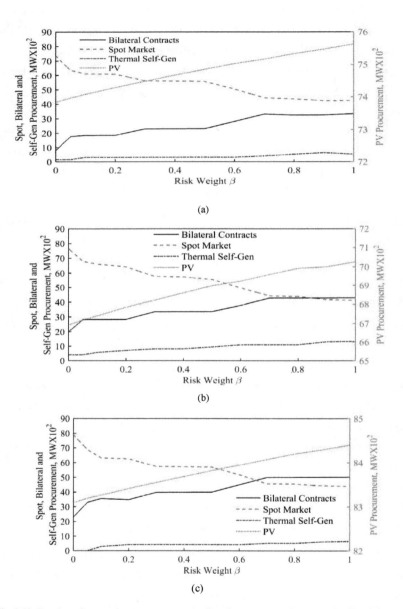

Fig. 3.12: Purchase from various procurement options for consumer classes (a). Residential, (b). Agricultural, and (c). Industrial

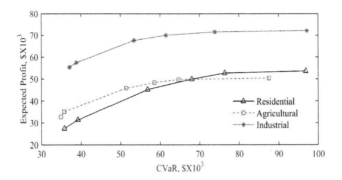

Fig. 3.13: Efficient frontier for various consumer classes

3.5 Conclusions

The work presented in this chapter determines optimal dynamic sale prices and optimal energy procurement portfolio targeting profit maximization of LSE under elastic consumer demand. Spot market price uncertainty is characterized and represented in the presented work. Further, impact **high RE penetration on LSE's decision making** is analyzed.

As observed from results, dynamic sale prices for three consumer classes shift demand from high price hours to low price hours. This **implies that LSE's costly energy** procurement is reduced in corresponding hours from spot market, thus reducing energy procurement cost. This reduction in energy procurement costs increases LSE's profit. During risk minimization, demand shifts from highly risky hours to less risky hours, which reduces price risk exposure in those hours for LSE. The impact of risk aversion parameter on procurement portfolio selection and sale prices are highlighted. Case study highlights that the same decision making strategies cannot be applied to different consumer classes to **maximize LSE's profit. However, prices follow spot market price** pattern. The presence of RE reduces the prices during RE availability hours, and this accumulates demand during RE availability hours. Due to high RE utilization, LSE procures less from other options, i.e., bilateral contracts, thermal self-generation, and spot markets. This **significantly increases LSE's expected** profit and minimizes spot market price risk. This model is suitable for an LSE providing services in a retail market where competitive forces are still silent and LSEs intend to offer new service, especially dynamic sale prices.

Consumer behavior modeling is an important aspect for LSE's profit maximization. In this chapter, consumer elastic behavior is modeled considering its elasticity. Consumers may rationally revise their consumption profiles in response to sale price signals to minimize purchase cost. This necessitates further investigation of consumer behavior modeling and change in strategies to maximize LSE's profit. Consumers with flexible demand interact with LSE in a hierarchical manner. Therefore, modeling hierarchical relationship between LSE and consumer is an important extension of proposed work presented in this chapter.

Chapter – 4: LSE's Risk Based Profit Maximization in Bi-level Framework

This chapter presents a mathematical model to capture interaction between LSE and consumer to maximize LSE's profit.

Chapter - 4
LSE's Risk Based Profit Maximization in Bi-level Framework

4.1 Introduction

Consumer's price responsive behavior influence LSE's sale price and energy procurement decisions [1], [6]. This consumer behavior is modeled through elasticity and demand function in Chapter 3. Elasticity indicates sensitivity of consumers to change in sale prices [12]. Hence, consumer's revised consumption is reflection of their sensitivity to sale price. However, consumer may rationally determine optimal energy consumption profile based on flexible demand and price signals received from LSE to minimize its cost [132]. Consumer's flexible demand could be different from elastic demand. Elastic demand is result of consumer's sensitivity to sale prices [12], whereas flexible demand is result of change in operations of shiftable/ interruptible/ curtailable load [133].

This economic optimization goal of consumer conflicts with LSE's objective [84], [132]. Moreover, relationship between consumer and LSE is regarded as hierarchical relationship [132]. Under this relationship, both LSE and consumer act as decision-makers. They make decisions to maximize or minimize their respective objective functions. However, decisions of both decision-makers are impacted by decision of others [134]. Therefore, existence of this hierarchical relationship necessitates its consideration in LSE's decision making to capture consumer behavior based on its economic goal rather than elasticity. Modeling of hierarchical relationship requires advanced mathematical model such as bi-level programming model [135].

This chapter explicitly presents a hierarchical decision making model to determine optimal dynamic sale prices and energy procurement decisions for an LSE. Instead of consumer's elastic demand, its flexible demand is considered. RE availability and spot market price uncertainty are considered on LSE's decision making. Bi-level programming is applied to model consumer's economic optimization and LSE's profit maximization objectives.

Chapter - 4

Considering medium-term planning of LSE, a small thermal generation unit and RE as self-generation, along with spot market and bilateral contracts, is considered for procuring electricity. Risk of spot market prices is modeled by CVaR approach to devise risk-based strategies. Formulated bi-level optimization problem is transformed into an equivalent single-level problem using the classical Karush-Kuhn-Tucker (KKT) approach. Obtained non-linear formulation is linearized by introducing auxiliary variables and duality theory. A case study on the PJM market is proposed to show the effectiveness of proposed approach. Results from the proposed model help LSE to analyze its procurement and selling price strategies when consumer makes consumption decision rationally based on its flexible demand.

4.2 Problem Description

LSE determines optimal energy procurement and dynamic sale prices to be offered to consumers in medium-term to maximize profit. It procures energy from WEM through bilateral contracts and spot market. A part of its energy requirement is procured from available thermal and RE self-generation. Uncertainty of spot market price introduces volatility in procurement cost, eventually making the profit risky [6]. Therefore, LSE aims to minimize profit-risk according to its risk aversion level. Hence, a risk-averse decision making model for LSE is considered. Further, consumers can adjust their consumption based on sale prices. This price-responsive behavior of consumers influences LSE's decision making problem. Consumers as decision-maker, optimize their consumption pattern to minimize energy procurement cost (energy bill) based on offered sale price dynamics [97]. Hence, the decision making problem consists of two decision-makers (LSE and consumers), each trying to optimize its respective objective (which are conflicting), subject to respective constraints. LSE's decision making is formulated as a bi-level programming problem, as illustrated in Fig. 4.1.

The upper-level of bi-level problem describes LSE's risk-based profit maximization. To analyze LSE's risk-averse behavior, trade-off between risk and profit for LSE has been modeled using CVaR framework. The lower-level of bi-level is consumers' cost minimization problem. Both LSE and consumers optimize their objectives over a jointly dependent set.

The bi-level optimization problem is solved by transforming into an equivalent single-level problem by replacing lower-level problem with KKT optimality conditions. The

KKT condition appears as Lagrangian and complementarity slackness constraints in equivalent single-level problem. The resulting problem is non-linear, which is linearized by introducing an auxiliary variable and strong duality theory.

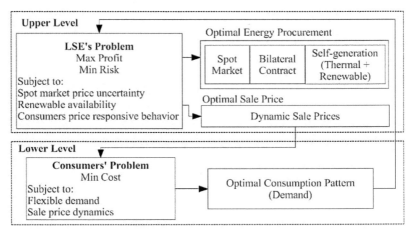

Fig. 4.1: Bi-level decision making problem

It is considered that the aggregated outcome from RE would be provided to LSE. Due to aggregation, RE generation uncertainty consideration is avoided [136]. However, the impact of RE variability and spot market prices is high; hence both are considered to determine sale prices. In the proposed problem, consumers are assumed to react rationally to information provided by LSE.

4.3 Mathematical Model

The proposed bi-level problem discussed in Section-4.2 is mathematically formulated in this section.

4.3.1 Upper Level: Load Serving Entity's Problem

LSE's trade-off, i.e., profit maximization and risk minimization, subject to procurement and sale price constraints, is mathematically formulated in the following subsections.

4.3.1.1 Procurement Cost

Total cost for energy procurement in hour t is equal to the sum of energy cost from bilateral contracts C_t^B, thermal generating units C_t^{TG}, renewable generation C_t^{PV}, and

Chapter - 4

spot market C_t^S.

$$Cost_t^{total} = C_t^B + C_t^{TG} + C_t^{PV} + C_t^S \tag{4.1}$$

Mathematical equations for bilateral contracts, thermal self-generation, and spot market procurement are similar to the work presented in Chapter 3. However, mathematical model of RE differs in this chapter. Hence, it is discussed in detail.

4.3.1.1.1 Renewable Generation

In this work, RE generation from PV system has must-run status, providing PV_t^{act} generation. Energy procurement cost from PV systems is considered using (4.2). This cost is accounted to recover investment cost of PV systems made by LSE. LSE incurs some cost when it procures RE through contracts, though fuel cost for RE generation is negligible. This necessitates consideration of RE cost. The value of λ_t^{PV} can be mathematically calculated using bottom-up or any other methodology by considering relevant cost parameters of PV systems. Calculation of λ_t^{PV} is beyond the scope of the work, so its value is assumed. Available RE generation could be curtailed during periods of low demand using (4.3).

$$C_t^{PV} = PV_t \lambda_t^{PV} \tag{4.2}$$

$$PV_t = PV_t^{act} + PV_t^{cur} \tag{4.3}$$

4.3.1.2 Energy Balance Constraint

Energy balance constraint is represented by constraint (4.4). Here L_t^{act} is consumers' optimized demand, obtained from the lower level problem.

$$P_t^S + \sum_{i \in I} P_{i,t}^B + P_t^{SG} = L_t^{act} \tag{4.4}$$

4.3.1.3 Revenue

LSE resells acquired energy to its consumers at a per unit sale prices λ_t^{sale}. Revenue for each time is determined by (4.5), subject to sale price constraints ((4.6)-(4.8)).

$$Revenue_t = L_t^{act} \lambda_t^{sale} \tag{4.5}$$

4.3.1.4 Sale Price Modeling

Constraints (4.6) to (4.8) represents sale price model. Constraints (4.6) and (4.7) restrict the sale price to go below and above a certain percentage of nominal sale price at period t. Constraint given by (4.8) is used to impose an upper bound on the average of the sale prices. This constraint ensures sufficient number of low sale prices hours. It also helps to alleviate LSE's market power [97], [132]. In the absence of (4.8), LSE's optimal choice will be maximum limit on sale price, i.e., $\lambda_t^{sale} = (1 + z_t^{max}) . \lambda_t^{nsale}$, which seems to be unfair to consumers. It is assumed that LSE and consumers have mutually agreed upon considered minimum, maximum, and average sale prices.

$$\lambda_t^{sale} \geq (1 - z_t^{min}) . \lambda_t^{nsale} \tag{4.6}$$

$$\lambda_t^{sale} \leq (1 + z_t^{max}) . \lambda_t^{nsale} \tag{4.7}$$

$$\frac{\sum_t L_t^{act} \lambda_t^{sale}}{\sum_t L_t^{act}} \leq \lambda^{avg} \tag{4.8}$$

4.3.1.5 Price Uncertainty Modeling

Spot market prices are uncertain and can not be predicted with certainty beforehand at the time of decision making. Uncertainty in spot market prices introduces risk for LSE [6]. This risk in LSE's energy procurement is quantified and measured using CVaR. CVaR is computed by taking a weighted average of the losses exceeding VaR values, as presented in Section 3.3.5 of Chapter 3. Risk incorporated by spot market procurement is accounted by (3.26).

4.3.1.6 Load Serving Entity's Objective Function

LSE's objective is to maximize expected profit and minimize risk imposed by spot market prices. LSE determines a trade-off between expected profit and risk considering its risk-preferences. This trade-off is formulated based on CVaR criterion in the upper level of bi-level problem. LSE's multi-objective function is formulated by combining profit and risk objectives using a risk weighting parameter β. The LSE's multi-objective function is given by

$$Maximize \quad Obj_{LSE} = \sum_t E(Prof_t) - \beta \sum_t Risk_t \tag{4.9}$$

Chapter - 4

$$\sum_t E(Prof_t) = \sum_t Revenue_t - \sum_t E(Cost_t) \tag{4.10}$$

The first term in (4.9), $E(Prof_t)$ represents LSE's expected profit and is obtained by subtraction of expected procurement cost (4.1) from revenue generated (4.5), which is given by (4.10). The second term in (4.9) represents risk, which is to be minimized. LSE's objective function (4.10) is subjected to procurement, sale price, and energy balance constraint.

4.3.2 Lower Level: Consumers' Problem

Consumer is second decision-maker of the proposed bi-level problem. Hence, at lower level, consumers determine their optimal energy consumption based on dynamic sale prices offered by LSE. Consumers received dynamic price signals from LSE. LSE determines these prices in the upper-level of the proposed bi-level problem. Group of consumers with similar characteristics can be represented by different classes. For these consumer classes, mathematical formulation remains same. However, available flexible demand and nominal demand profile of these consumer classes will be different. Therefore, this is reflected in the mathematical model through appropriate factors. Moreover, consumer is aware about its nominal/ initial demand profile and available flexible demand.

4.3.2.1 Consumers' Objective Function

Consumer objective of energy bill minimization is mathematically formulated as

$$Minimize \quad Obj_{con} = \sum_t L_t^{act} \lambda_t^{sale} \tag{4.11}$$

subject to constraints

$$L_t^{act} \geq L_t x^l \tag{4.12}$$

$$L_t^{act} \leq L_t x^u \tag{4.13}$$

In (4.11), L_t^{act} is revised energy demand, which is a decision variable for consumer of considered class. Inequality constraints (4.12) and (4.13) describe the minimum and maximum limits on consumers' demand in terms of its initial demand [12]. These constraints represent relation between revised demand, initial demand, and available flexible demand of consumer. The minimum and maximum limits on energy consumption are determined by multiplying a factor x^l and x^u on initial demand, respectively. These factors represent consumers' flexible demand in percentage. As it is mentioned before, consumers have information about their initial demand profile and flexible demand. Consumers aim to redistribute their demand to minimize energy bill. Hence, this work considers that consumer does not reduce or suppress its total demand. However, they can shift it to other hours while maintaining the total demand to original value. In other words, sum of L_t over 24 hours is equal to the sum of L_t^{act}. This demand shift is mathematically formulated as (4.14).

$$\sum_t L_t^{act} = \sum_t L_t \qquad (4.14)$$

4.3.3 Equivalent Single-Level Bi-Level Model

Bi-level problem formulated for LSE can be solved by transforming it to one level. This transformation is done by adding equivalent KKT optimality conditions [137] of lower-level problems (4.11)-(4.14) as constraints to the upper-level problem. This includes both necessary and sufficient conditions for optimality [137]. Resultant problem contains complementary and slackness conditions, making it non-linear. The non-linear complementary conditions are replaced by its equivalent linear expressions to make the problem linear.

4.3.3.1 KKT Optimality Condition for Lower Level Problem

The KKT optimality conditions are obtained by determining Lagrangian functions for lower-level problem (4.11)-(4.14), given by (4.15) [96], [138]. σ, γ_t, and μ_t in (4.15), represent the Lagrange multiplier associated with the lower-level problem constraints (4.12)-(4.14), respectively.

$$L_a = \sum_t L_t^{act} \lambda_t^{sale} - \sigma \left[\sum_t L_t^{act} - \sum_t L_t \right] - \sum_t \gamma_t \left[L_t^{act} - L_t x^l \right] - \sum_t \mu_t \left[L_t x^u - L_t^{act} \right] \qquad (4.15)$$

KKT necessary optimality conditions of lower-level problem are obtained by partial derivatives of Lagrangian function (4.15). KKT necessary optimality conditions consist of stationarity (4.16), dual feasibility (4.17)-(4.18), and complementary slackness (4.19)-(4.20) conditions, respectively. These conditions are in addition to the primal feasibility constraints (4.12)-(4.14). Therefore, lower level problem is represented by (4.12)-(4.14) and (4.19)-(4.20). To this end, bi-level optimization problem is

Chapter - 4

transformed into a non-linear complementary model.

$$\frac{\partial L_a}{\partial L_t^{act}} = \lambda_t^{sale} - \sigma - \gamma_t + \mu_t = 0 \qquad (4.16)$$

$$\gamma_t \geq 0 \qquad (4.17)$$

$$\mu_t \geq 0 \qquad (4.18)$$

$$\gamma_t \left[L_t^{act} - L_t . x^l \right] = 0 \qquad (4.19)$$

$$\mu_t \left[L_t . x^u - L_t^{act} \right] = 0 \qquad (4.20)$$

4.3.3.2 Equivalent Linear Expression

The nonlinear complementary slackness conditions (4.19)-(4.20) obtained from KKT conditions are linearized by introducing an auxiliary binary variable (w_t and υ_t) and a sufficiently positive large constant (M_1 and M_2) [137]. This results in two constraints per complementarity constraints as

$$\gamma_t \leq M_1 w_t \qquad (4.21)$$

$$\left[L_t^{act} - L_t . x^l \right] \leq M_1 (1 - w_t) \qquad (4.22)$$

$$w_t \in \{0, 1\} \qquad (4.23)$$

$$\mu_t \leq M_2 \upsilon_t \qquad (4.24)$$

$$\left[L_t . x^u - L_t^{act} \right] \leq M_2 (1 - \upsilon_t) \qquad (4.25)$$

$$\upsilon_t \in \{0, 1\} \qquad (4.26)$$

Constraints (4.21)-(4.23) replace (4.19) and constraints (4.24)-(4.26) replace (4.20). Product of consumer demand and sale prices, i.e., $L_t^{act} \lambda_t^{sale}$ in Eq. (4.11) introduces non-linearity in the problem. Its equivalent linear expression is determined by utilizing duality theory. The dual of lower-level problem (4.11)-(4.14) is obtained and shown in (4.27).

$$Maximize \quad \sigma \sum_t L_t + \sum_t \gamma_t L_t x^l - \sum_t \mu_t L_t x^u \qquad (4.27)$$

78

where σ, γ_t, and μ_t are the dual variables associated with the constraints (4.12)-(4.14). Optimal solution is obtained by equating primal (4.11) and dual (4.27) as

$$\sum_t Revenue_t = \sum_t L_t^{act} \lambda_t^{sale} = \sigma \sum_t L_t + \sum_t \gamma_t L_t x^l - \sum_t \mu_t L_t x^u \qquad (4.28)$$

It is worth to point out here that (4.28) does not connect lower and upper level. Non-linear term $Revenue_t = L_t^{act} \lambda_t^{sale}$ in objective function (4.9) is replaced by its equivalent linear-term represented in (4.28). Eq. (4.10) can be rewritten as

$$\sum_t E(Prof_t) = \sigma \sum_t L_t + \sum_t \gamma_t L_t x^l - \sum_t \mu_t L_t x^u - \sum_t E(Cost_t) \qquad (4.29)$$

4.3.4 Overall Objective Function

Objective function of equivalent single level problem is obtained by substituting (4.29) in (4.9) as

$$Maximize \quad Obj_{LSE} = \sigma \sum_t L_t + \sum_t \gamma_t L_t x^l - \sum_t \mu_t L_t x^u - \sum_t E(Cost_t) - \beta \sum_t Risk_t \qquad (4.30)$$

The equivalent single-level formulation of the bi-level problem consists of is a Mixed Integer Linear Programming (MILP) problem. Objective function (4.30) is to be maximized subject to procurement, sale price, energy balance, CVaR, consumer constraints, and KKT optimality conditions.

4.4   Case Study

4.4.1 Data

The proposed formulation for an LSE is illustrated via a case study. Historical spot market price data from the PJM electricity market is considered for this study [128]. RE generation profile for a solar based PV system in the Pennsylvania region [139] is considered (shown in Fig. 4.2). Fixed price of 38 $/MWh is considered for energy procurement from RE. It is assumed that one type of thermal self-generating unit is available with LSE. Specification for considered thermal generating unit is shown in Table 3.1. One bilateral contract at 35 $/MWh is considered with maximum and minimum procurement limits of 300 MW and 30 MW, respectively. Dynamic sale prices and energy procurement are determined for each hour of a day for LSE.

Chapter - 4

This case study considers one type of consumer class to illustrate performance of proposed model. The model can be applied to other consumer classes with their respective demand profiles and flexibility. Forecasted demand profile of considered consumer class is depicted in Fig. 4.3. It is considered that consumers can vary their demand up to 15% of their original demand at each hour. Nominal sale price for each hour is set equal to 5% higher than corresponding spot market prices. An average sale price of 38 $/MWh is considered. To keep sale prices within a reasonable limit, LSE considers minimum value of sale prices equal to nominal sale prices and maximum sale price 20% higher than the nominal sale price at each hour.

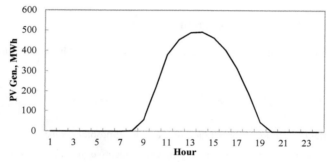

Fig. 4.2: PV generation profile

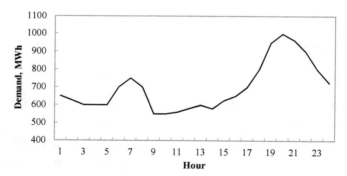

Fig. 4.3: Consumer demand profile

4.4.2 Statistical Calculations

Expected values of spot market prices for each hour are calculated as mean of historical data (from 1/12/2016 to 11/03/2017) corresponding to that hour, over the entire planning period. Value of CVaR for 95% confidence level is computed for per unit value of historical spot market price data. Historical simulation method, as described in

Chapter 3, is utilized to obtain per unit CVaR values. Calculated expected spot market prices and CVaR values over 24 hours are similar to as shown in Fig. 3.4 of Chapter 3.

4.4.3 Simulations

The objective function (4.30) subject to procurement, sale price, energy balance, consumer, and KKT constraints is simulated for different values of risk weighting parameter β to maximize LSE's profit. The resulting MILP problem has been solved using CPLEX solver commercially available in GAMS® software on Intel®, Core(TM), i5 CPU, 3.20 GHz, and 4GB of RAM system [130]. CPLEX is a high-performance mathematical programming solver for MILP [131]. It uses branch and cut algorithm for solving MILP problems. Solution of the proposed problem generates 508 variables and 96 discrete variables. Average execution and computation time to solve the optimization problem is 0.089 s and 2.88 s, respectively.

4.4.4 Results

Considering LSE's risk preferences, i.e., risk-neutral $(\beta = 0)$ and risk-averse $(\beta > 0)$, results evaluated from simulations are illustrated through Figs. 4.4-4.8. Highest risk-averse behavior of LSE is obtained for $\beta = 1.5$. Results include dynamic sale prices, optimized consumer demand, and energy procurement portfolio.

4.4.4.1 Dynamic Sale Prices and Consumer Demand

Dynamics of optimized sale prices and flexible consumers' demand is shown in Figs. 4.4 and 4.5, respectively, for risk-neutral and risk-averse LSE. Curves for sale price and demand labeled with $\beta = 0$ and $\beta = 1.5$ represent risk-neutral and risk-averse behavior of LSE. Dynamics of consumers' optimized demand indicates that consumers modify their energy consumption during each period to minimize their overall energy cost (Figs. 4.4 and 4.5). In other words, consumers' response follows the sale price signal communicated by the LSE, which provides them minimum cost (energy bill). It shifts most of the flexible demand to periods (hours) when sale prices are low. This demand shift also benefits LSE to procure most of its demand during low market price hours.

Fig. 4.4 shows that the dynamics of offered sale prices for risk-neutral and risk-averse case differs. This difference in sale price dynamics is governed by wholesale prices and spot market price risk. As risk-neutral LSE ignores spot market price risk, it offers

Chapter - 4

higher value of sale prices to hours when procurement cost tends to be high. However, a risk-averse LSE considers a given level of spot market price risk and tries to offer high sale prices during hours where spot market price risk is high. In both cases, LSE does this to shift its consumers' flexible demand from these hours to other beneficial hours. Consumers respond to these prices, and this results in different consumption pattern for risk-neutral and risk-averse cases (Fig. 4.5). At certain hours, sale prices are low, though consumers reduce their demand (Hours 20, 21 for risk-neutral case). This happens because consumers can benefit by shifting demand to other hours when sale prices are comparatively low (Hours 15-17) (Figs. 4.4 and 4.5).

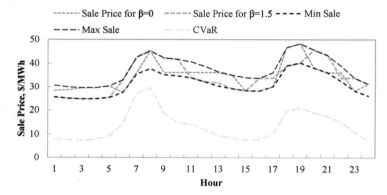

Fig. 4.4: Dynamics of sale prices

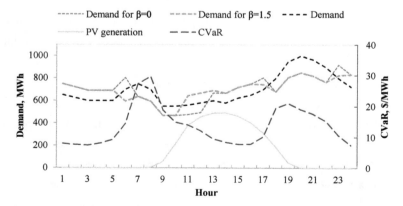

Fig. 4.5: Dynamics of consumers' optimized demand

4.4.4.2 Energy Procurement

Energy procurement from the spot market, bilateral contracts, and thermal and RE self-generation obtained for risk-neutral and risk-averse LSE is shown in Figs. 4.6 and 4.7, respectively. These energy allocations are determined for respective consumers' optimized demand (Fig. 4.5).

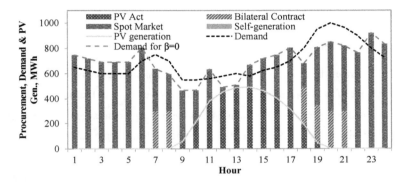

Fig. 4.6: Procurement from various options for risk-neutral LSE

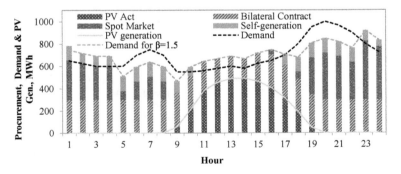

Fig. 4.7: Procurement from various options for risk-averse LSE

Figs. 4.6 and 4.7 show that risk-neutral LSE procures most of its energy requirement from spot markets, whereas risk-averse LSE procures most of its energy from sources with low uncertainty. It can be observed from obtained energy allocation that available RE generation is completely utilized in both cases. This utilization significantly impacts energy procurement from cost-effective sources during its availability. To decrease overall procurement cost, risk-neutral LSE tries to increase consumers' demand during RE availability hours when spot market prices are low (Fig. 4.6). For instance, at Hours

13-17, consumer demand increases, which is procured from spot market as prices are low. To increase consumer demand in these hours, LSE offers lower sale prices to consumers (Fig. 4.4). Risk-averse LSE also tries to increase consumers' demand during RE availability hours to minimize risk imposed by uncertain spot market prices. This happens to increase procurement from low uncertainty sources, which was low during RE availability (Fig. 4.7). LSE offers lower sale prices during RE availability and increases with high risk (Fig. 4.4).

4.4.4.3 Impact of Risk

Fig. 4.8 illustrates the efficient frontier depicting the risk-averse behavior of LSE for different value of β. Frontier implies that for a high value of β, LSE's expected profit and associated risk decreases. Depending upon the willingness to accept spot market price risk, LSE may select decisions corresponding to that. In case value of β is too high, (following the pattern of the frontier, that can be implicated) it would reduce risk and expected profit to zero, which indicates that LSE's participation in retail activities is not recommended. This means risk-averse LSE can reduce risk up to a certain level.

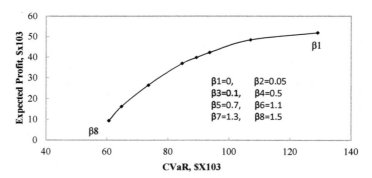

Fig. 4.8: Efficient frontier

4.5 Conclusions

This chapter models hierarchical interaction between LSE and consumers. Optimal dynamic sale prices and energy procurement decisions are determined subject to spot market price uncertainty and RE availability. Consumer's optimal consumption profile is obtained to minimize its energy bill. The hierarchical relationship is explicitly modeled using bi-level programming. Hence, LSE's objective of profit maximization and risk minimization is employed at upper level. Consumers' objective of optimizing cost for determining its consumption profile is employed in the lower level. Resultant bi-level problem is transformed into single level using KKT optimality conditions. LSE's risk-neutral and risk-averse behavior is explicitly analyzed in hierarchical framework.

Case study shows that a risk-neutral LSE maximizes its profit by providing price incentives (low sale prices) to consumers to shift their demand to low spot market prices duration. However, risk-averse LSE provides price incentives when risk is low. Sale price dynamics is also impacted by RE utilization. Both risk-neutral and risk-averse LSE offer sale prices to increase consumers' demand during RE availability to utilize RE generation. However, as risk averseness increases, LSE shifts its procurement from risky spot market to risk-free contracts. Thus, LSE effectively utilizes RE by active use of consumers' flexible demand to achieve its objective.

This chapter contributes by presenting a different approach for LSE's profit maximization. Modeling of rational consumer and its interaction with LSE is highlighted. This modeling differs from consumer behavior model presented in previous Chapter 3, which was based on consumer elasticity.

In Chapters 3 and 4, a significant emphasis is laid on consumer behavior modeling through elasticity and modeling hierarchical relationship. CVaR approach is considered to determine risk-based procurement and sale price decisions for LSE. However, CVaR approach is quite suitable when sufficient information on uncertain parameter (spot market price) is available. In the absence of sufficient information/ data, decisions are made by considering certain assumptions on missing information/ data. This may lead to imprudent decisions, and as a consequence, LSE exposes itself to unexpected financial losses. This necessitates investigation of LSE's profit maximization, considering a robust approach. State-of-the-art on uncertainty modeling presented in Chapter 2, implies that IGDT is a suitable approach for such cases. In the next chapter, IGDT framework is considered to determine and analyze LSE's robust decision making for its profit maximization.

Chapter – 5: LSE's Risk Based Profit Maximization under Severe Uncertainty

This chapter presents IGDT based risk based profit maximization approach to model favorable and unfavorable face of severe uncertainty.

Chapter - 5
LSE's Risk Based Profit Maximization under Severe Uncertainty

5.1 Introduction

Impact of spot market price uncertainty on LSE's profit is addressed using CVaR in Chapters 3 and 4. However, when LSE is subject to severe uncertainty or under lack of sufficient information about uncertain parameters, conventional uncertainty modeling approaches cannot be applied [29], [63]. Conventional approaches rely on probability distribution or membership function of uncertain parameters [6], [59]. These functions cannot be extracted or determined with insufficient information of uncertain parameters [29], [30]. Hence, certain assumptions are made to determine the function for uncertain parameters. Under such conditions, robustness of decisions cannot be guaranteed. Moreover, decisions can be imprudent and may lead to unexpected outcomes such as high losses to decision-makers. Hence, effective uncertainty modeling approaches are required to model severe uncertainty.

In Chapters 3 and 4, adverse/ unfavorable face of spot market price uncertainty that may incur losses to LSE is modeled and analyzed. However, uncertainty can be advantageous/ favorable at various instances that may provide windfall gain [29]. LSE can exploit opportunity of favorable spot market price uncertainty to maximize its profit. As CVaR can quantify only losses due to adverse/ unfavorable uncertainty, an alternative uncertainty modeling approach is required to devise optimal decisions for favorable uncertainty. State-of-the-art review presented in Chapter 2 advocates IGDT approach for modeling severe uncertainty, and devising adverse and favorable decisions for decision-makers. Hence, IGDT framework is used for LSE's profit maximization.

In the previous chapters, spot market, bilateral contracts, and self-generation are mainly considered as energy procurement sources. Evolving storage technologies such as BES can help LSE to devise new strategies for maximizing its profit. ESS may enhance flexibility to manage LSE's procurement portfolio and consumer demand [40], [140]. This requires strategic or smart charge/ discharge of ESS to maximize profit and minimize risk of LSE. LSE's decision making with ESS is focused in this chapter. This

Chapter - 5

chapter extends work addressed in Chapter 3 to comprehend LSE's decision making with ESS under IGDT framework.

This chapter proposes IGDT based model to determine dynamic sale prices to maximize LSE's profit under severe spot market price uncertainty. Adverse and favorable conditions arising due to varying spot market prices are modeled for robustness and opportunistic decision making to address risk-averse and risk-seeking behavior of LSE. Consumers' price sensitive elastic demand is considered via their price elasticity. Consumer's concern of increase in energy bill is modeled in sale price model. In addition to spot market, bilateral contracts, and self-generation, ESS is considered to enhance flexibility for managing procurement decisions and consumer demand. LSE's procurement and dynamic sale price decisions are analyzed under severe spot market price uncertainty. Effectiveness of proposed model is highlighted based on correlation between demand and spot market price profile. Results from the simulations indicate that proposed model ensures consumer benefit and increases LSE's profit.

5.2   Problem Description

This work considers problem of setting/ optimizing dynamics of sale prices for an LSE aiming to maximize its profit, focusing on consumer's price sensitivity to change in sale prices. Optimized sale price dynamics encourage consumers to redistribute their demand by shifting loads to more favorable hours (periods with low sale prices), increasing profit of LSE. LSE is not interested to decrease total energy consumed by consumers, but in redistributing energy between hours. Based on redistributed load (demand) profile, LSE decides the quantum of energy to be procured from various available resources. LSE makes sale price and procurement decisions well in advance with deficit/ imperfect information about future spot market prices. This imprudent decision making due to lack of information about future spot market prices may lead to significant losses.

As discussed in previous chapters, LSE procures energy from spot market, bilateral contracts, and thermal and RE based self-generation facility. EES is additional source considered for procuring a portion of total LSE's demand. RE generation is a must-run generation. Therefore, its availability has significant impact on dynamic sale prices and procurement from other sources. Bilateral contracts are fixed price contracts, and self-generation (thermal and RE) cost is known at the time of decision making. IGDT based model is proposed to set/ optimize dynamic sale price and procurement decisions, simultaneously. Spot market price uncertainty is modeled using an envelop-bound info-gap model of uncertainty. Historical data of spot market prices is considered without relying on any distribution function in the uncertainty model, and to devise robust and opportunity decisions. LSE's risk-averse and risk-seeking behavior is addressed by modeling its profit maximization problem using robustness and opportunity functions, respectively. Impact of RE availability and EES operation on sale price dynamics has been investigated under the said scenario.

5.3   Mathematical Model

LSE's decision making problem is modeled as explained in Section 5.2. LSE's profit maximization is modeled without considering spot market price uncertainty. This determines maximum profit that can be attained by LSE. Then, IGDT based formulation is presented for risk based profit maximization of LSE.

5.3.1 Procurement Cost

Procurement cost incurred from bilateral contracts C_t^B, thermal self-generation C_t^{TG}, RE self-generation C_t^{PV}, and spot market C_t^S is calculated as

$$Cost_t^{total} = C_t^B + C_t^{TG} + C_t^{PV} + C_t^S \quad or \quad Cost_t^{total} = C_t^B + C_t^{SG} + C_t^S \tag{5.1}$$

Mathematical constraints of these procurement sources are same as considered in Chapter 3.

5.3.1.1 Energy Storage System

In this work, ESS is considered as an additional source for energy procurement. ESS is an effective source to manage energy portfolio due to its fast dispatch capability. However, efficiency of ESS depends on its type. ESS based on battery, hydrogen fuel, and pump hydro, etc. can be utilized by LSE. For this work, Battery Energy Storage (BES) system is considered as a part of procurement portfolio. However, a different ESS based on alternate technology can be considered, if the aim is to analyze the impact of ESS technology types on LSE's decisions. Cost of BES is neglected as focus of work is not to analyze the problem from investment perspective. This work assumes that LSE

Chapter - 5

may recover investment cost by charging a minimal fixed amount to consumers, in addition to its energy bill. Battery operation is bound by following constraints

$$v_{k,t}^c (P_e^c)^{min} \leq P_{e,t}^c \leq v_{e,t}^c (P_e^c)^{max}; \quad v_{e,t}^d (P_e^d)^{min} \leq P_{e,t}^d \leq v_{e,t}^d (P_e^d)^{max} \tag{5.2}$$

$$SOC_e^{min} \leq SOC_{e,t} \leq SOC_e^{max} \tag{5.3}$$

$$SOC_{e,t} = SOC_{e,t-1} + \left(\eta_e^c P_{e,t}^c - P_{e,t}^d / \eta_e^d \right) \tag{5.4}$$

$$v_{e,t}^c + v_{e,t}^d \leq 1; \quad v_{e,t}^c, v_{e,t}^d \in \{0,1\} \tag{5.5}$$

Charging and discharging power of battery e is limited by (5.2). State of Charge (SOC) of battery e is limited by constraint (5.3). Battery energy dynamics model is given by (5.4). Battery charging/ discharging status is modeled to avoid simultaneous charging and discharging of battery e (5.5).

5.3.2 Energy Balance Constraint

As BES is considered as one of procurement options in addition to bilateral contracts, self-generation, and spot market, energy balance constraint is modified, and given as

$$P_t^S + \sum_{i \in I} P_{i,t}^B + \sum_{g \in G} P_{g,t}^{TG} + P_t^{RE} + \sum_{e \in E} P_{e,t}^c = L_t^{act} + \sum_{e \in E} P_{e,t}^d \tag{5.6}$$

5.3.3 Revenue

Revenue generated by LSE is given by (5.7). It depends on offered sale prices. L_t^{act} is consumers price responsive demand. This is obtained as shown in Section 3.3.4 of Chapter 3.

$$Revenue = L_t^{act} \lambda_t^{sale} \tag{5.7}$$

5.3.4 Sale Price Modeling

Mathematical model for sale price determination is presented in Chapter 3. To restrict consumer's energy bill for an increase, an additional constraint (5.8) on sale price is considered. This constraint is considered in sale price model of Chapter 4. Constraint (5.8) indicates that average of offered sale prices over a day cannot exceed a predetermined average value. The value of average price could be flat price or a price on which consumer and LSE agree.

92

$$\frac{\sum_t L_t^{act} \lambda_t^{sale}}{\sum_t L_t^{act}} \leq \lambda^{avg} \tag{5.8}$$

5.3.5 Objective Function

LSE's objective function for profit maximization is given by (5.9). It is obtained by subtracting procurement cost (5.1) from revenue (5.7).

$$
\begin{aligned}
Prof &= \sum_t Revenue_t - \sum_t Cost_t^{total} \\
&= \sum_t L_t^{act} \lambda_t^{sale} - \left(\sum_t P_t^S \bar{\lambda}_t^S + \sum_t \sum_{i \in I} P_{i,t}^B \lambda_{i,t}^B + \sum_t C_t^{SG} \right)
\end{aligned} \tag{5.9}
$$

Equation (5.1) neglects spot market price uncertainty. Evaluated profit indicates maximum profit that can be attained by LSE. This formulation indicates LSE's risk-neutral behavior. In this work, uncertainty is modeled using IGDT. IGDT and mathematical formulation based on it are presented in next sections, respectively.

5.4 Information Gap Decision Theory Method

IGDT supports a decision-maker to evaluate robust decisions in an uncertain environment. These decisions guarantee a specified target (value) provided that actual value of uncertain parameter falls into its maximized uncertainty horizons centered on the given forecast. IGDT-based decisions consider available information about uncertain parameters. Uncertainty is modeled as a variation interval between what is known and what could be known [29]. IGDT maximizes the horizon of uncertainty and provides a solution that guarantees a certain expectation for the objective. For example, a decision-maker is exposed to uncertain prices and desires a guaranteed profit for which it intends to know values of the decision variables. It is assumed that a set of price forecasts for the next day is available to decision-maker. An IGDT-based solution guarantees a specified critical profit, provided the after-the-fact prices fall into a maximized price band centered at the forecast prices (see Fig. 5.1). This band, in general, is sometimes referred to as robustness region.

System model, uncertainty model, and performance requirement are three components of IGDT method [29].

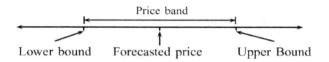

Fig. 5.1: Illustration of information gap uncertainty

1) System Model:

   System model describes the input/ output structure of system under consideration. System model is interpreted as the reward of decision maker for selected values of decision variable when considering the uncertain parameter.

2) Uncertainty Model:

   Uncertainty model embodies information about uncertain parameters. Uncertain parameter is represented by an unbounded family of nested sets centered around its forecasted value. An uncertainty model is used to model the gap between the known and what needs to be known. As discussed in Chapter 2, there are various info-gap uncertainty models such as Envelope-bound, Energy-bound, Minkowski-norm, Slope-bound, and Fourier-bound, *etc* [29]. Depending upon the requirement, appropriate model can be selected.

3) Performance Requirement:

   A performance model is a function of decision variable and uncertainty parameter. Uncertain parameter variations can either be adverse or favorable. Based on decision maker's risk management strategy, two performance functions can be defined in an info-gap model. A risk-averse decision maker desires to make decisions in a way to be immune against effects of unfavorable deviations of the uncertain parameter from its corresponding forecast value. For this, a robustness function is modeled, which provides immunity against unfavorable deviations of uncertain parameters. A risk-seeker decision maker will make decision in a way to benefit from favorable deviations of the uncertain parameter from the corresponding forecasted value. An opportunity function models the possible high profits for the risk-seeker decision-maker.

5.5 Load Serving Entity's Decision Making: Information Gap Decision Theory Framework

IGDT assesses performance requirements for decisions based on targeted/ anticipated outcomes by robustness and opportuneness functions addressing risk-averse or risk-seeking behavior of decision-maker, respectively. Immunity from unfavorable movements and opportunity of windfall gain for favorable movements of spot market prices are determined using robustness and opportuneness functions, respectively. The mathematical model based on IGDT approach is presented in the following sub-sections.

5.5.1 Decision Variables

Uncertainty horizon α, spot market prices λ_t^S, sale price λ_t^{sale}, revised demand L_t^{act}, and quantum of energy to be procured from procurement options (spot market P_t^S, bilateral contracts $P_{i,t}^B$, self-generation P_t^{SG}, and charging power $P_{e,t}^c$, and discharging power $P_{e,t}^d$ of battery) are decision variables of the problem, represented as $Q = \left[\alpha, \lambda_t^S, \lambda_t^{sale}, L_t^{act}, P_t^S, P_{i,t}^B, P_t^{SG}, P_{e,t}^c, P_{e,t}^d \right]$.

5.5.2 Price Uncertainty Modeling

Info-gap models uncertainty by the unbounded family of nested sets, which are nested around expected value of the parameter of interest. Each set represents a particular degree of knowledge deficiency, depending on the level of nesting [29]. In this work, envelop-bound info-gap model is considered to model spot market price uncertainty. This uncertainty model does not hold any assumption on the probability distribution, which makes it suitable for the cases with a high level of uncertainty or lack of sufficient information. In this model, magnitude of deviation is proportional to the forecasted value. Mathematically, model is given as [29]

$$U(\alpha, \tilde{\lambda}_t) = \left\{ \left| \frac{\lambda_t - \tilde{\lambda}_t}{\tilde{\lambda}_t} \right| \le \alpha; \alpha \ge 0 \right\} \tag{5.10}$$

where , α is uncertainty of uncertain parameter λ_t, $\tilde{\lambda}_t$ is forecasted/predicted value of λ_t and $U(\alpha, \tilde{\lambda}_t)$ is set of all values of λ_t whose deviation from $\tilde{\lambda}_t$ will never be more than $\alpha\tilde{\lambda}_t$.

Chapter - 5

5.5.3 Performance Requirement

Performance requirement is the minimum requirement which is expected or anticipated from the system, under severe uncertainty. For this requirement, decision variables are evaluated based on robustness and opportuneness functions. Robustness function captures the adverse face of the uncertainty, whereas the opportuneness function takes advantage of favorable situations.

Robustness function provides the highest level of uncertainty for which minimum requirements are always fulfilled. For LSE, robustness guarantees that the profit would always be higher than targeted critical profit π_c. Robustness represents risk-averse behavior of decision-maker by providing immunity against losses under maximum uncertainty, defined by uncertainty horizon α. Mathematically, robustness function $\hat{\alpha}(\pi_c)$ can be expressed as

$$\hat{\alpha}(\pi_c) = \max_{\alpha} \left\{ \alpha : \min_{\lambda_t \in U(\alpha,\tilde{\lambda}_t)} \pi(Q,\lambda_t) \geq \pi_c \right\} \tag{5.11}$$

Opportuneness function evaluates the possibility of achieving higher profit resulting from favorable face of uncertainty and provides windfall benefits. It addresses risk-seeking attitude of decision-makers. The opportuneness function expresses the least level of uncertainty (i.e., uncertainty horizon α) for which windfall profit can be achieved greater than anticipated profit π_w. Mathematically, opportuneness function $\hat{\beta}(\pi_w)$ can be expressed as

$$\hat{\beta}(\pi_w) = \min_{\alpha} \left\{ \alpha : \max_{\lambda_t \in U(\alpha,\tilde{\lambda}_t)} \pi(Q,\lambda_t) \geq \pi_w \right\} \tag{5.12}$$

5.5.3.1 Robustness Function

Robustness function targets to immune the LSE from the possibility of losses or low profit attributable to undesired deviations in spot market prices from the forecasted value. $\hat{\alpha}(\pi_c)$ for LSE represents the largest value of uncertainty, so that minimum profit is not less than desired/ targeted profit π_c. LSE's profit is given by (5.9). Therefore, from (5.9) and (5.11), robustness function is mathematically expressed as

$$\max_{\alpha} \left[\min \left[\sum_t L_t^{act} \lambda_t^{sale} - \left(\sum_t P_t^S \tilde{\lambda}_t^S + \sum_t \sum_{i \in I} P_{i,t}^B \lambda_{i,t}^B + \sum_t C_t^{SG} \right) \right] \geq \pi_c \right] \tag{5.13}$$

With the given uncertainty α, minimum profit would occur when spot market prices are highest. Therefore, according to (5.10), λ_t^S is replaced with $\tilde{\lambda}_t^S(1+\alpha)$ to obtain robustness function. This is expressed as

$$\hat{\alpha}(\pi_c) = \max\left[\frac{\sum_t L_t^{act}\lambda_t^{sale} - \sum_t P_t^S \tilde{\lambda}_t^S - \sum_t\sum_{i\in I} P_{i,t}^B \lambda_{i,t}^B - \sum_t C_t^{SG} - \pi_c}{\sum_t P_t^S \tilde{\lambda}_t^S}\right] \tag{5.14}$$

The risk level of LSE's planning can be controlled by changing targeted profit, π_c, which is decided by LSE according to its risk-attitude. Here, the value of π_c indicates risk-averseness of LSE.

5.5.3.2 Opportunity Function

A risk-seeking LSE can take advantage of the favorable face of uncertainty to attain windfall gain. Opportunity function evaluates the least level of uncertainty so that profit can be as large as π_w. Therefore, from (5.9) and (5.12), opportunity function is mathematically expressed as

$$\min_{\alpha}\left[\max\left[\sum_t L_t^{act}\lambda_t^{sale} - \left(\sum_t P_t^S \tilde{\lambda}_t^S + \sum_t\sum_{i\in I} P_{i,t}^B \lambda_{i,t}^B + \sum_t C_t^{SG}\right)\right] \geq \pi_w\right] \tag{5.15}$$

The maximum profit can occur when spot market prices are lowest. Therefore, according to (5.10), λ_t^S is replaced with $\tilde{\lambda}_t^S(1-\alpha)$ to obtain opportuneness function. This is expressed as

$$\hat{\beta}(\pi_w) = \min\left[\frac{\pi_w - \sum_t L_t^{act}\lambda_t^{sale} + \sum_t P_t^S \tilde{\lambda}_t^S + \sum_t\sum_{i\in I} P_{i,t}^B \lambda_{i,t}^B + \sum_t C_t^{SG}}{\sum_t P_t^S \tilde{\lambda}_t^S}\right] \tag{5.16}$$

Derived robustness function (5.14) has to be maximized, whereas opportuneness function (5.16) has to be minimized subject to energy procurement, sale price, and energy balance constraints. The values of π_c and π_w are determined by profit maximization problem posed by (5.9), subject to energy procurement, sale price, and energy balance constraints.

5.6 Case Study

Proposed model is illustrated for cases with and without RE and BES. Results for risk-neutral, robustness, and opportuneness for each case are analyzed and presented in the following sub-sections.

5.6.1 Data

This study considers historical spot market prices data of the PJM electricity market [128]. Predicted future spot market prices are shown in Fig. 5.2. Bilateral contracts at fixed price of 32 $/MWh with maximum and minimum trading limits of 460 MW and 30 MW, respectively, are considered. Specification of BES parameters is shown in Table-5.1. Renewable generation availability from solar is shown in Fig. 5.3 [139]. Solar energy price is presumed at 15$/MWh.

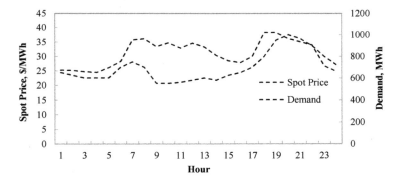

Fig. 5.2: Spot market price and demand profile

Table 5.1: Specification of battery energy storage system

Parameter	Value	Parameter	Value
SOC_0	100 MW	$\bar{p}^c$	$0.2\, SOC^{max}$
SOC^{max}	300 MW	$\underline{p}^c$	0
$(P^d)^{max}$	$0.2\, SOC^{max}$	η_c	95%
$(P^d)^{min}$	0	η_d	90%

This case study considers residential consumer class. However, analysis for other consumer classes can be conducted by using relevant information. Considered demand profile of residential consumer is depicted in Fig. 5.2. This demand profile is assumed based on the nature of usual electricity consumption patterns. Value of price elasticity is considered at -0.8 [87]. Flat prices are considered higher than from average of expected spot market prices, equal to 38 $/MWh. The calculated average value of WEM equals 31 $/MWh.

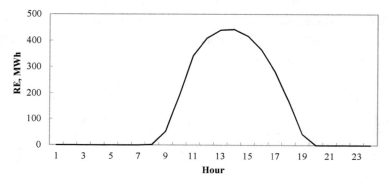

Fig. 5.3: Renewable profile

5.6.2 Simulations

The objective function (5.9) is simulated subject to energy procurement, sale price, and energy balance constraints. This simulation provides maximum possible value of profit $\tilde{\pi}(\tilde{\lambda}^S)$, for projected spot market prices $\tilde{\lambda}_t^S$ for each case. Evaluated decisions without considering uncertainty imply risk-neutral behavior of LSE. Considering maximum profit $\tilde{\pi}(\tilde{\lambda}^S)$, number of critical profit targets π_c and anticipated windfall gains π_w are assumed. π_c and π_w are values of profit considered less than and higher than $\tilde{\pi}(\tilde{\lambda}^S)$, respectively.

For each value of π_c and π_w, robustness (5.14) and opportuneness (5.16) optimization problems are simulated subject to energy procurement, sale price, and energy balance constraints, respectively, to determine decision variables Q. These three MINLP optimization problems are solved using SBB/ CONOPT solver under GAMS [130]. SBB uses Standard B&B algorithm for node selections. This solution is utilized by

Chapter - 5

CONOPT to optimize NLP [131]. These simulations are carried out on an Intel®
Core™ i7, 2 GHz processor, and 8 GB RAM system.

5.6.3 Results

Maximum profit $\bar{\pi}(\bar{L}^{act}, \tilde{\lambda}^S)$ obtained from simulation for risk-neutral with and without
RE and BES system is 182477\$ and 129824\$, respectively. Energy procurement from
spot market, bilateral contracts, self-generations, and BES system are shown in Table-
5.2. Hourly dynamic sale prices and revised demand obtained are depicted in Fig. 5.4.

Based on the maximum profit value, values of π_c varying from 182477\$ to 132296\$
and 129824\$ to 94122\$ are considered for case with RE and BES and without RE and
BES, respectively. Values of π_w that vary from 182477\$ to 222621\$ and 129824\$ to
158385\$ are considered for cases with RE and BES and without RE and BES,
respectively. Results obtained for robustness and opportunity functions are shown in
Figs. 5.5 to 5.9. Curve on left-hand side of the marker indicates outcomes related to
robustness and the right is related to opportuneness.

Table 5.2: Energy procurement from various sources

	With RE and BES (MW)	Without RE and BES (MW)
Bilateral Contract	4014.84	5520.00
Spot Market	9214.78	10820.00
Thermal Self-Generation	420.00	420.00
RE	3161.26	-
BES	300.00	-

5.6.3.1 Robust Strategy

Robustness and opportuneness curves show that determined decisions can tolerate
higher spot market price uncertainty as critical profit target π_c reduces (Fig. 5.5).
Hence, high robustness is obtained at low profit. To make decisions robust, impact of
spot market price uncertainty is reduced by decreasing energy procurement from spot
market and increasing it from bilateral contracts and thermal-generation (Fig. 5.5a). As
RE is must-run generation, its procurement remains fixed (Fig. 5.5b). However, BES
operation is adjusted considering spot market energy procurement to increase

robustness of decisions and optimize procurement cost. This strategy increases expected energy procurement cost, and thus, low expected profit is achieved for high robustness. This increase in cost is borne by LSE to attain robustness. Higher robustness indicates higher energy cost. Fig. 5.6a represents robust energy procurement decisions when RE and BES are part of energy procurement in LSE's decision making process. Fig. 5.7a represents robust energy procurement decisions when RE and BES are not a part of energy procurement.

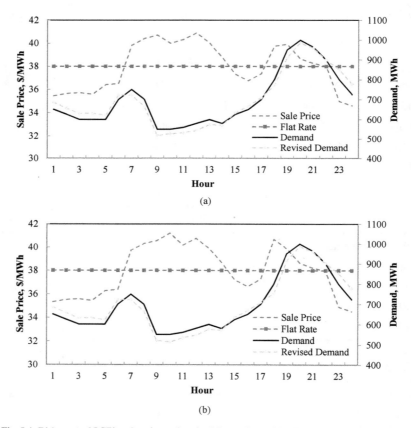

Fig. 5.4: Risk-neutral LSE's sale price and revised demand (a) with RE and BES and (b) without RE and BES

Chapter - 5

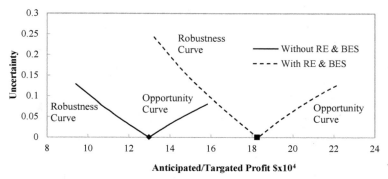

Fig. 5.5: Robustness and opportuneness curves for critical and anticipated target

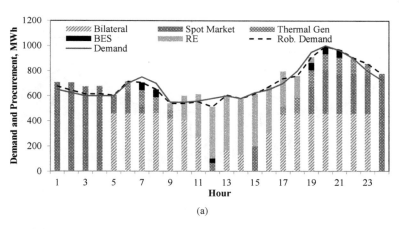

(a)

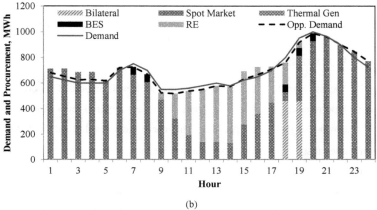

(b)

Fig. 5.6: Energy procurement from different sources of LSE with RE and BES for (a) Robustness and (b) Opportuneness

LSE's Risk Based Profit Maximization under Severe Uncertainty

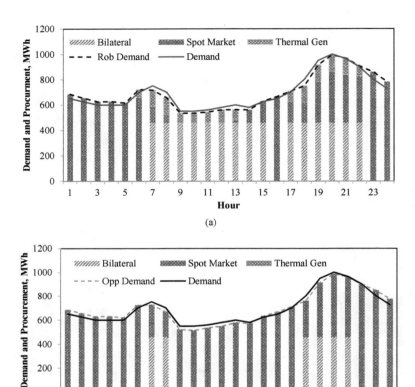

Fig. 5.7: Energy procurement from different sources of LSE without RE and BES for (a) Robustness and (b) Opportuneness

5.6.3.2 Opportunistic Strategy

In Fig. 5.5, the curve on the right side of the marker indicates outcome of opportuneness function. Opportuneness function determines the lowest required level of uncertainty α at which windfall gain π_w is possible. This curve indicates that anticipated profit target increases as spot market price uncertainty increases. Energy procurement strategy is shown in Figs. 5.6b and 5.7b for case with and without RE and BES, respectively. It shows that energy procurement from spot market increases and procurement from bilateral contracts and self-generation decreases with increase in π_w. This change in spot market share is required to increase price variability to achieve anticipated profit.

Chapter - 5

As spot market trading has high price variations, hence it has high windfall probability. Fig. 5.6b indicates that BES operation is changed in opportuneness case to achieve anticipated profit.

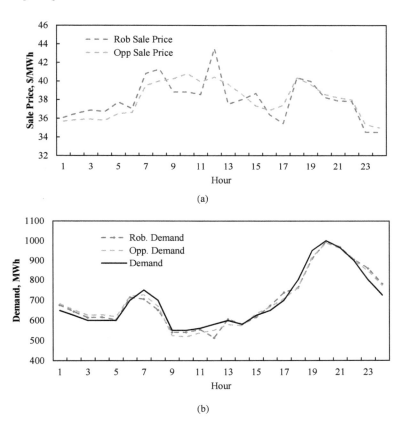

Fig. 5.8: Sale prices and demand of LSE with RE and BES for (a) Robustness and (b) Opportuneness

5.6.3.3 Sale Prices and Demand

Dynamic sale prices and revised demand obtained for robustness and opportuneness function for LSE with and without RE and BES are shown in Figs. 5.8 and 5.9, respectively. Offered sale prices follow spot market prices in those hours where LSE procures energy from spot market (Figs. 5.8a and 5.9a). However, when procurement from spot market is nil at Hours 8-11, 13-14, and 16-17, sale prices follow the price of fixed-price contracts (Figs. 5.6a and 5.8a). A peak at Hour 12 for robust price reduces demand to reduce spot market procurement (Fig. 5.6a and 5.8). Presence of RE and BES allows offering lower sale prices (Figs. 5.8a and 5.9a).

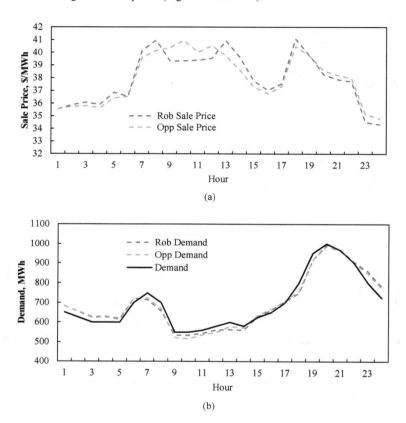

Fig. 5.9: Sale prices and demand of LSE without RE and BES for (a) Robustness and (b) Opportuneness

Correlation factor is determined between demand and spot market price profiles for cases with and without RE and BES. Reduced value of correlation factor (Table-5.3) indicates that the offered dynamic sale prices modify the demand in such a way that it does not follow spot market prices. This avoids high cost of procurement from markets. A lower value of correlation between (revised) demand and spot market price profile compared to base case justifies shift in consumer demand from high to low price hours. Here, base case indicates correlation between initial demand and spot market price profile.

Chapter - 5

Table 5.3: Correlation between demand and spot market prices

	Without RE	With RE
Base Case	0.4395	0.4395
Risk-Neutral	0.2324	0.2498
Robustness	0.2597	0.2658
Opportuneness	0.2945	0.2598

5.6.3.4 Consumer Energy Cost

Table-5.4 indicates consumer cost obtained under flat and dynamic pricing. Offered dynamic sale prices minimize consumer cost (payment) in both cases.

Table 5.4: Consumer cost under flat and dynamic pricing

Cost in $	Flat Rate	Dynamic Prices (without RE)	Dynamic Prices (with RE)
Risk-Neutral	636880	635364	635522
Robustness	636880	635506	635372
Opportuneness	636880	635598	635731

5.7　Conclusions

This chapter proposes an IGDT based LSE's profit maximization model considering spot market price uncertainty. RE and BES system are considered as procurement options along with spot market, bilateral contracts, and thermal generation. Spot market price uncertainty is modeled using envelop bound info-gap model. Risk-averse and risk-seeking behavior of LSE is modeled by robustness and opportuneness functions.

The proposed model is illustrated via cases with and without RE and BES. Results indicate that for a risk-averse LSE, its decision becomes robust with lower profit targets as tolerance for uncertainty increases. To enable the opportunity of high windfall profits, larger uncertainties are required. Offered sale prices shift consumer's demand to low sale price periods that reduce consumers' cost and increases LSE profit. A reduced correlation factor between demand and spot market price profile implies shift in consumer demand. BES scheduling to high spot market price hours reduces procurement cost and increases robustness of decisions.

This chapter exclusively models severe spot market price uncertainty under IGDT framework. As LSE decision making is also affected by uncertain demand, its modeling becomes essential particularly when a strong correlation exists between uncertain demand and prices. IGDT has capability to model multiple uncertainties with relationship among these uncertain parameters. The next chapter presents a novel decision making framework for LSE's profit maximization based on correlated demand and spot market prices.

Chapter – 6: LSE's Risk Based Profit Maximization Framework for Correlated Uncertainties

This chapter describes correlated spot market price and demand modeling under IGDT framework to maximize LSE's profit.

Chapter - 6

LSE's Risk Based Profit Maximization Framework for Correlated Uncertainties

6.1 Introduction

In previous chapters, LSE profit maximization strategies are derived considering spot market price uncertainty. Spot market price uncertainty is highly uncertain and therefore, emphasis on its modeling is given. In these chapters, consumer demand is considered varying and its impact of demand shifting is analyzed for dynamic sale prices. As discussed in Chapter 2, consumer demand and spot market prices are correlated due to their inherent dependence [2], [6]. Together demand and spot market price uncertainty can create severe situations for LSE, and it may face extreme losses. Such extreme risk may arise due to significant co-variation of demand and spot market prices.

Consideration of correlated variations of demand and spot market prices can improve LSE's profit margin [6]. However, modeling of this correlation is challenging. This correlation can be modeled by stochastic framework using scenario tree approach [31]. However, scenario tree cannot model extreme situations (spikes) and suffers when strong correlation exists between modeled variables. This may lead to imprudent decisions. Such issues with existing frameworks necessitate a novel framework to model demand and spot market price uncertainties, along with their correlation.

As discussed in Chapter 2, IGDT framework can be considered to model severe multiple uncertainties [30]. IGDT has been applied with integral modeling of demand and spot market price uncertainties for LSE. Envelop bound info-gap model can model these uncertainties one at a time to develop robust strategies. In particular, demand and spot market price uncertainties are considered by multi-objective Pareto optimal solutions [63], [71]. However, existing research do not model demand and spot market price uncertainties in a single framework, hence are unable to consider co-variations existing between them. This makes existing LSE's profit maximization decision making modeling incomplete.

Chapter - 6

This chapter proposes a novel framework to determine LSE's energy procurement and sale price decisions targeting maximum profit under demand and spot market price uncertainty, using IGDT. Available information of consumer demand and spot market prices, along with their correlation, is used in Ellipsoid-bound info-gap uncertainty model. Adverse and favorable conditions arising due to varying consumer demand and spot market prices are modeled to take robust and opportunistic decisions. This way, IGDT addresses risk-averse and risk-seeking behavior of LSE. Impact of multiple severe uncertainties has been highlighted for LSE's electricity procurement and sale price, separately for weekdays and weekends based on information from historical data of considered case study. Results from case study indicate that capturing demand and price uncertainties together improves LSE's decision making.

6.2   Problem Description

This work intends to determine a large LSE's procurement and sale price decisions considering integrated modeling of correlated demand and spot market prices. LSE decides sale prices to be offered to supply its varying consumer demand and procurement from spot market at varying prices. It plans these decisions in advance with deficit information about future demand and spot market prices. As the quantum of electricity to be procured and spot market prices are uncertain, and can not be known perfectly at the time of decision making, imprudent decision making may lead to significant losses, especially when both demand and spot market prices are high.

As aligned with previous work in this thesis, LSE fulfills its requirement through procurement from WEM via spot market, bilateral contracts, and self-generation facility. Market liquidity is assumed to be sufficient. A framework is proposed to model both spot market prices and demand uncertainties with their correlated variations using ellipsoid bound model of uncertainty under IGDT framework. Risk-averse and risk-seeking behavior of LSE is addressed by modeling its profit maximization problem using robustness and opportunity functions, respectively.

6.3   Mathematical Model

The LSE's profit maximization problem is mathematically formulated with two components: procurement cost and revenue, which are described without considering price and demand uncertainty. This formulation determines maximum profit that can be attained by LSE. Then IGDT based framework is presented to model correlated price and demand uncertainty.

6.3.1 Procurement Cost

LSE's procurement cost at hour t is the sum of the cost of energy procured from bilateral contracts C_t^B, self-generation C_t^{SG}, and spot market C_t^S represented as

$$Cost_t^{total} = C_t^B + C_t^{SG} + C_t^S \tag{6.1}$$

Mathematical formulation and associated constraints for bilateral contracts, thermal self-generation, and spot market are similar to Chapter-3.

6.3.2 Energy Balance Constraint

Energy balance constraint is given as

$$P_t^S + \sum_{i \in I} P_{i,t}^B + P_t^{SG} = L_t \tag{6.2}$$

6.3.3 Revenue

LSE generates revenue by selling energy to its consumers at offered sale prices λ_t^{sale}. Revenue for each time slot is

$$Revenue_t = L_t \lambda_t^{sale} \tag{6.3}$$

6.3.4 Sale Price Modeling

Sale price is modeled by following constraints

$$\lambda_t^{min} \leq \lambda_t^{sale} \leq \lambda_t^{max} \tag{6.4}$$

$$\frac{\sum_t L_t \lambda_t^{sale}}{\sum_t L_t} \leq \lambda^{avg} \tag{6.5}$$

Constraint (6.4) restricts LSE's sale prices within the minimum and maximum values of sale prices. Inequality constraint (6.5) is used as a bound on the average sale prices. This constraint ensures sufficient number of low price periods. In the absence of constraint (6.5), LSE would always charge the maximum price to consumer when not

Chapter - 6

faced by high regulation price. It is assumed that LSE and consumer will agree on minimum, maximum, and average value of sale prices.

6.3.5 Objective Function

LSE's expected profit (6.6), is calculated by subtracting procurement cost (6.1) from generated revenue (6.3). Eq. (6.7) is obtained by substituting values of revenue and cost term in (6.6).

$$Prof = \pi(L, \lambda) = \sum_t Revenue_t - \sum_t Cost_t^{total} \tag{6.6}$$

$$Prof = \pi(L, \lambda) = \sum_t L_t \lambda_t^{sale} - \left(\sum_t P_t^S \lambda_t^S + \sum_t \sum_{i \in I} P_{i,t}^B \lambda_{i,t}^B + \sum_t C_t^{SG} \right) \tag{6.7}$$

LSE's profit function given by (6.6) (or (6.7)) evaluates maximum profit of LSE. This is the profit that can be attained by LSE when demand and spot market uncertainties are neglected. This function represents LSE's risk-neutral behavior. As demand and spot market price uncertainties and their correlation affect LSE's profit, these uncertainties are modeled using IGDT model. This is presented in subsequent section.

6.4 Load Serving Entity's Decision Making: Information Gap Decision Theory Framework

In this work, IGDT models demand and spot market price uncertainties and their co-variability. Modeling of uncertain demand and spot market prices in a single framework is proposed by considering a single uncertainty horizon, under Ellipsoid-bound uncertainty model. Immunity from unfavorable movements and opportunity of windfall gain for favorable movements of demand and spot market prices are determined by formulating robustness and opportuneness formulations, respectively. The mathematical model based on IGDT approach is presented in following sub-sections.

6.4.1 Price and Demand Uncertainty Modeling

Ellipsoid bound uncertainty model of info-gap approach is used to model uncertainty of total demand to be procured L_t and spot market prices λ_t^S, at hour t. This model considers uncertainty by an unbounded family of nested sets, nested around expected value of the parameter of interest. Each set represents a particular degree of knowledge deficiency, depending on the level of nesting [29]. Ellipsoid is an envelope of uncertainty, centered around best estimate vector $\tilde{x}_t$, with a distance represented by uncertainty horizon α. Here, x_t is a vector representing uncertain parameters, i.e., demand and spot market prices as shown in (6.8). In real-time, actual values of uncertain parameters may vary from their estimates $\tilde{x}_t$ in either direction (positive or negative) by an error Δx_t (See (6.9)-(6.11)).

$$x_t = [L_t, \lambda_t^S] \tag{6.8}$$

$$x_t = \tilde{x}_t + \Delta x_t \tag{6.9}$$

i.e., $$[L_t, \lambda_t^S] = [\tilde{L}_t + \Delta L_t, \tilde{\lambda}_t^S + \Delta \lambda_t^S] \tag{6.10}$$

$$\Delta x_t = [\Delta L_t, \Delta \lambda_t^S] \tag{6.11}$$

Ellipsoid quantifies the uncertainty by measuring distance between best estimate and reality (actual value) as shown in Fig. 6.1. Larger distance indicates larger uncertainty α (Fig. 6.1). The shape of ellipsoid describes the relative degree of variability of demand and spot market prices and their correlated behavior. A variance-covariance matrix W defines this information. Here, for the case of two uncertain parameters, L_t and λ_t^S, it is a 2X2 matrix for each interval t. Its elements, w_{11} and w_{22} are variance of L_t and λ_t^S, respectively, and $w_{12} = w_{21}$ are covariances between demand and spot market price. This information is obtained from historical data. Mathematically, ellipsoid bound info-gap model is given in (6.12) [29]

$$U(\alpha, x_t) = \{x_t : \Delta x_t W^{-1} \Delta x_t' \le \alpha^2\}, \ \alpha \ge 0 \tag{6.12}$$

Here $'$ represents transpose.

6.4.2 Performance Requirement

Performance requirement is the minimum requirement expected or anticipated from the system, under severe uncertainty. Robustness and opportuneness functions are formulated to attain targeted profit π_c and anticipated profit π_w, respectively. Mathematically, robustness function $\hat{\alpha}(\pi_c)$ and opportuneness function $\hat{\beta}(\pi_w)$ can be expressed by (6.13) and (6.14), respectively.

$$\hat{\alpha}(\pi_c) = \max_{\alpha}\{\alpha : \min \pi(Q, x) \ge \pi_c\} \tag{6.13}$$

$$\hat{\beta}(\pi_w) = \min_{\alpha}\{\alpha : \max \pi(Q, x) \geq \pi_w\} \tag{6.14}$$

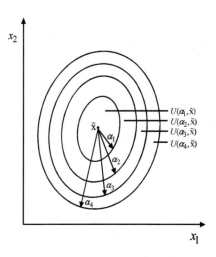

Fig. 6.1: Pictorial representation of Ellipsoid bound info-gap model

6.4.2.1 Robustness Function

Robustness function defines adverse face of uncertainty and targets to immune the decision-maker from possibility of losses. $\hat{\alpha}(\pi_c)$ for LSE represents the largest value of uncertainty, so that minimum profit is not less than desired/ targeted profit π_c. Minimum profit is determined by substituting (6.10) and (6.12) in (6.7). This would yield

$$\min_{\Delta x_t} Prof = \sum_t (\tilde{L}_t + \Delta L_t) \lambda_t^{sale} - \left(\sum_t P_t^S (\tilde{\lambda}_t^S + \Delta \lambda_t^S) + \sum_t \sum_{i \in I} P_{i,t}^B \lambda_{i,t}^B + \sum_t C_t^{SG} \right) \tag{6.15}$$

s.t. $\Delta x_t W^{-1} \Delta x_t' \leq \alpha^2$ \hfill (6.16)

Equation (6.15) and (6.16) can be simplified as

$$\min_{\Delta x_t} Prof =$$
$$\sum_t \tilde{L}_t \lambda_t^{sale} - \left(\sum_t P_t^S \tilde{\lambda}_t^S + \sum_t \sum_{i \in I} P_{i,t}^B \lambda_{i,t}^B + \sum_t C_t^{SG} \right) + \min_{\Delta x_t} \left\{ \sum_t \Delta L_t \lambda_t^{sale} - P_t^S \Delta \lambda_t^S : \Delta x_t W^{-1} \Delta x_t' \leq \alpha^2 \right\}$$
\hfill (6.17)

Substituting $y_t = [\lambda_t^{sale} - P_t^S]$ in (6.17) as

$$\min_{\Delta x_t} Prof = \sum_t \tilde{L}_t \lambda_t^{sale} - \left(\sum_t P_t^S \tilde{\lambda}_t^S + \sum_t \sum_{i \in I} P_{i,t}^B \lambda_{i,t}^B + \sum_t C_t^{SG} \right) + \min_{\Delta x_t} \left\{ \sum_t \Delta x_t y_t' : \Delta x_t W^{-1} \Delta x_t' \leq \alpha^2 \right\}$$

$$(6.18)$$

Using Lagrangian relaxation method for the convex optimization problem, first-order optimality condition is obtained as

$$\nabla_{\Delta x_t, \mu} \left\{ \Delta x_t y_t' + \mu (\alpha^2 - \Delta x_t W^{-1} \Delta x_t') \right\} = 0 \qquad (6.19)$$

where μ is a Lagrange multiplier. Derivatives of (6.19) results in

$$y_t' - 2\mu W_t^{-1} \Delta x_t' = 0 \qquad (6.20)$$

$$\alpha^2 - \Delta x_t W_t^{-1} \Delta x_t' = 0 \qquad (6.21)$$

$$\Delta x_t' = \frac{1}{2\mu} W_t y_t' \text{ and } \alpha^2 = \Delta x_t W_t^{-1} \Delta x_t' \qquad (6.22)$$

Simplifying (6.22) would result in

$$\alpha^2 = \frac{1}{2\mu} W_t y_t W_t^{-1} \frac{1}{2\mu} W_t y_t' = \frac{1}{4\mu^2} y_t W_t y_t' \qquad (6.23)$$

$$\frac{1}{2\mu} = \pm \frac{\alpha}{\sqrt{y_t W_t y_t'}} \qquad (6.24)$$

$$\Delta x_t = \pm \alpha \frac{y_t W_t}{\sqrt{y_t W_t y_t'}} \qquad (6.25)$$

$$\Delta x_t y_t' = \pm \alpha \sqrt{y_t W_t y_t'} \qquad (6.26)$$

Considering minimum value from (6.26), (6.18) can be rewritten as

$$\min_{\Delta x_t} Prof = \sum_t \tilde{L}_t \lambda_t^{sale} - \left(\sum_t P_t^S \tilde{\lambda}_t^S + \sum_t \sum_{i \in B} P_{i,t}^B \lambda_{i,t}^B + \sum_t C_t^{SG} \right) - \left\{ \alpha \sum_t \sqrt{y_t W_t y_t'} \right\} \qquad (6.27)$$

As per robustness function defined in (6.13), the minimum profit should be at least equal to π_c. Hence,

$$\sum_t \tilde{L}_t \lambda_t^{sale} - \left(\sum_t P_t^S \tilde{\lambda}_t^S + \sum_t \sum_{i \in I} P_{i,t}^B \lambda_{i,t}^B + \sum_t C_t^{SG} \right) - \left\{ \alpha \sum_t \sqrt{y_t W_t y_t'} \right\} = \pi_c \qquad (6.28)$$

The uncertainty horizon at π_c can be defined as

Chapter - 6

$$\alpha(\pi_c) = \frac{\left\{\sum_t \tilde{L}_t \lambda_t^{sale} - \left(\sum_t P_t^S \lambda_t^S + \sum_t \sum_{i \in I} P_{i,t}^B \lambda_{i,t}^B + \sum_t C_t^{SG}\right)\right\} - \pi_c}{\sum_t \sqrt{y_t W_t y_t}{}'} \tag{6.29}$$

So the maximum value of uncertainty is

$$\hat{\alpha}(\pi_c) = \max_{Q_t} \frac{\left\{\sum_t \tilde{L}_t \lambda_t^{sale} - \left(\sum_t P_t^S \tilde{\lambda}_t^S + \sum_t \sum_{i \in B} P_{i,t}^B \lambda_{i,t}^B + \sum_t C_t^{SG}\right)\right\} - \pi_c}{\sum_t \sqrt{y_t W_t y_t}{}'} \tag{6.30}$$

6.4.2.2 Opportunity Function

A risk-seeking LSE evaluates opportunity function to take advantage of the favorable face of uncertainty and attain windfall gain π_w. As per Eq. (6.14), opportunity function (6.31) can be derived by substituting maximum value of profit and equating it to π_w.

$$\sum_t \tilde{L}_t \lambda_t^{sale} - \left(\sum_t P_t^S \tilde{\lambda}_t^S + \sum_t \sum_{i \in I} P_{i,t}^B \lambda_{i,t}^B + \sum_t C_t^{SG}\right) + \left\{\alpha \sum_t \sqrt{y_t W_t y_t}{}'\right\} = \pi_w \tag{6.31}$$

Value of uncertainty horizon α at anticipated windfall profit π_w is defined as

$$\beta(\pi_w) = \frac{\pi_w - \left\{\sum_t \tilde{L}_t \lambda_t^{sale} - \left(\sum_t P_t^S \tilde{\lambda}_t^S + \sum_t \sum_{i \in I} P_{i,t}^B \lambda_{i,t}^B + \sum_t C_t^{SG}\right)\right\}}{\sum_t \sqrt{y_t W_t y_t}{}'} \tag{6.32}$$

As opportuneness is minimum uncertainty requirement to obtain profit as large as π_w, it can be defined as

$$\hat{\beta}(\pi_w) = \min_{Q_t} \frac{\pi_w - \left\{\sum_t \tilde{L}_t \lambda_t^{sale} - \left(\sum_t P_t^S \tilde{\lambda}_t^S + \sum_t \sum_{i \in B} P_{i,t}^B \lambda_{i,t}^B + \sum_t C_t^{SG}\right)\right\}}{\sum_t \sqrt{y_t W_t y_t}{}'} \tag{6.33}$$

Derived robustness function (6.30) has to be maximized, whereas opportuneness function (6.33) has to be minimized subject to procurement, sale prices, and energy balance constraints. The values of π_c and π_w are determined by profit maximization problem posed by (6.7).

6.5 Case Study

The proposed model for a large LSE is illustrated via a case study. As demand and spot market prices are different for weekends and weekdays, their impact would be different on LSE's decisions. Hence, two sub-cases are formed considering weekend and weekdays to analyze LSE's decision making under correlated spot market price and demand uncertainty. Considering a planning period of one day and each hour as a sale price time block, robust and opportunity strategy is determined and presented.

6.5.1 Data

Historical demand and spot market price data from Nord Pool of West Denmark area is considered for this study [141]. This data is used to produce predicted values of demand and spot market prices for weekend and weekday (Fig. 6.2). Bilateral contracts at fixed price of 31.5 $/MWh and 39 $/MWh with maximum trading limits of 1300 MW and 1550 MW are considered for weekend and weekday respectively. A minimum limit of 10 MW on bilateral contracts is considered for both cases. LSE also has a self-generation facility of capacity 130 MW. The specification of thermal generating unit is same as specified in Chapter 3. Average sale prices of 40.5$ and 34.5$ for weekday and weekend are considered.

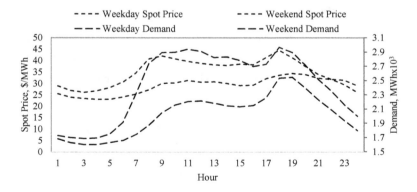

Fig. 6.2: Demand and spot market price curve for weekday (Mon.) and weekend (Sun.)

6.5.2 Statistical Calculations

Statistical calculations on historical demand and spot market price data are performed to obtain variance-covariance matrix W. This statistical analysis is performed for each day of a week over a season. A positive correlation between demand and spot market price is found for weekdays, whereas varying positive and negative correlation is observed for weekends. Negative correlation indicates that demand and spot market prices are changing in opposite direction. Based on these observations, Monday from weekday, and Sunday from the weekend, are selected for case study to analyze impact of positive and negative correlation on LSE's decision making. A representation of covariances in terms of correlation between spot market price and demand is shown in Fig. 6.3 for weekend and weekdays.

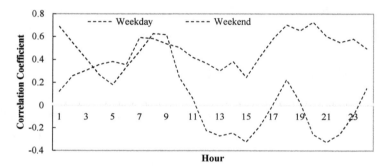

Fig. 6.3: Correlation between spot market price and demand for weekday (Mon.) and weekend (Sun.)

6.5.3 Simulations

The profit maximization problem is simulated for Eq. (6.7), subject to procurement, sale, and energy balance constraints, for the considered data. Based on the predicted value of spot market price $\tilde{\lambda}_t^S$ and demand $\tilde{L}_t$, optimization provides a maximum possible value of profit $\tilde{\pi}(\tilde{L},\tilde{\lambda}^S)$. This case evaluates decisions for risk-neutral LSE. Considering this maximum profit $\tilde{\pi}(\tilde{L},\tilde{\lambda}^S)$, critical profit targets π_c and anticipated windfall gains π_w are assumed in small steps. For robustness function, critical profit targets are considered less than $\tilde{\pi}(\tilde{L},\tilde{\lambda}^S)$. For opportunity function, anticipated windfall profits are considered higher than $\tilde{\pi}(\tilde{L},\tilde{\lambda}^S)$.

For each value of critical profit π_c and windfall gain π_w, decision variable Q_t is obtained by optimizing (6.30) and (6.33) subject to procurement, sale, and energy balance constraints. All the simulations and analysis are performed separately for weekdays and weekends.

The three optimization problems are MINLP in nature and solved using SBB/ CONOPT solver under GAMS [130]. The proposed model for LSE has 265 real and 48 discrete variables. Simulations are carried out on an Intel® Core™ i7, 2 GHz processor, and 8 GB of RAM system. Average execution and computation time to solve robust optimization problems for robustness are 0.253s and 10.31s and opportuneness are 0.3622s and 10.067s, respectively.

6.5.4 Results

The risk-neutral maximum value of profit $\tilde{\pi}(\tilde{L},\tilde{\lambda}^S)$ for weekday and weekend obtained by simulations are 304679.45$ and 263705.97$, respectively. This value is the maximum possible profit that can be optimally gained by LSE. The relevant quantum of energy procured from the spot, bilateral contracts, and self-generation are shown in Table 6.1. Hourly sale prices obtained are depicted in Fig. 6.4.

Table 6.1: Energy procurement from various sources

Day	Bilateral Contracts (MW)	Spot Market (MW)	Self-generation (MW)
Weekday	10850	51047.92	796.08
Weekend	7800	42522	0

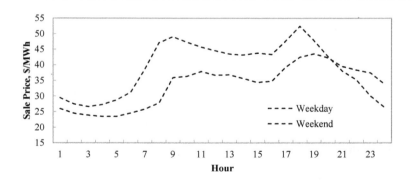

Fig. 6.4: Sale prices for risk-neutral LSE

For this case study, values of π_c varying from 304679.45 to 225462.79$ and from 263705.97 to 195142.42$ for weekdays and weekends, respectively, are considered. Values of π_w that vary from 304679.45 to 457019.19$ and from 263705.97 to 421929.55$ for weekday and weekends, respectively, are considered. The results obtained from simulations for each value of π_c and π_w for robustness and opportunity functions are shown collectively in Figs. 6.5 to 6.8. Curve on left-hand side of the marker indicates outcomes related to robustness and the right ones are related to opportuneness.

6.5.4.1 Robust Strategy

Results show that as value of critical profit target π_c reduces, decision making can tolerate higher uncertainties of spot market price and demand (Fig. 6.5). Hence, robustness of decision increases. In order to make robust decisions, LSE tries to avoid uncertainty and increases energy procurement from bilateral contracts and self-generation (Fig. 6.6). Procurement from spot market decreases to attain high robustness. This results in high expected energy procurement cost, and thus, expected profit reduces. The variation between expected cost and targeted profit for different values of robustness for weekday and weekend are shown in Fig. 6.7. Decrease in values of expected profit describes the cost for attaining robustness.

Fig. 6.5: Robustness and opportuneness curves for critical and anticipated target

6.5.4.2 Opportunistic Strategy

Right side of marker in Fig. 6.5 depicts the opportuneness curve. Opportuneness function determines the lowest required level of uncertainty α at which windfall gain π_w is possible. A beneficial uncertainty (defined on the y-axis of Fig. 6.5) is required to attain anticipated windfall gains (defined on the x-axis of Fig. 6.5). In case of opportuneness, value of uncertainty increases as anticipated profit target increases. Energy procurement from the spot, bilateral contracts, and self-generation for weekday and weekend are depicted in Fig. 6.6. It is observed that energy procurement from spot market increases through procurement from bilateral contracts and self-generation decreases with increasing anticipated profit π_w. This happens because high variability is desired to achieve anticipated profit. As spot (market) trading has high price variations, hence it possesses high windfall possibility.

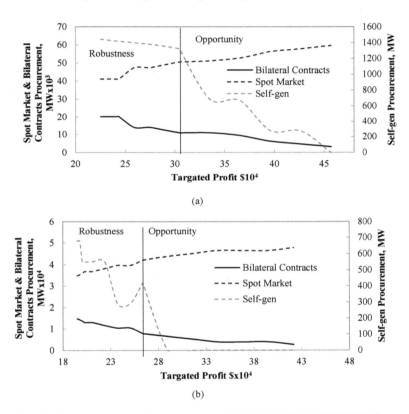

Fig. 6.6: Energy procurement from various sources for (a) Weekday and (b) Weekend

Fig. 6.7 gives expected cost and Fig. 6.8 depicts expected value of profit for weekday and weekend. These indicate expected costs and profit when demand/ spot market prices are equal to their forecasts. This indicates the cost of uncertainty, which represents that if anticipated deviation would not happen, expected profit will be lower and cost would be higher than its maximum/ minimum values.

For robustness and opportuneness, the dynamics of sale prices offered to the consumers are indicated in Fig. 6.9a. For robustness, high sale prices are offered during peak hours

Chapter - 6

to cope with increased cost experienced due to increased procurement from fixed cost sources (bilateral contracts and self-generation). However, low sale prices are offered for opportunity case during peak hours since procurement from spot market increases to exploit opportunity through uncertainty. Low volatility in spot market prices is experienced during Hours 10 to 16; hence, high and low sale prices are offered for opportuneness and robustness, respectively. This observation indicates that contrasting sale prices are consequences of LSE's risk-averse and risk-seeking behavior.

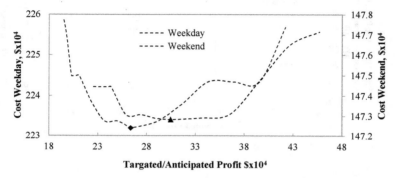

Fig. 6.7: Expected cost for weekend and weekday

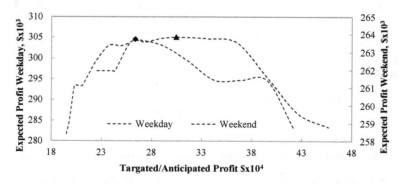

Fig. 6.8: Expected profit for various targeted profit for critical and anticipated target

The dynamics of sale prices offered for robustness (or opportuneness) case in weekend (Fig. 6.9b) differs from that is offered for weekday (Fig. 6.9a). Correlation between spot market price and demand characteristics impacts sale prices obtained for robustness and opportuneness. It is to be noted that correlation is positive for weekdays over 24 hours while for weekend it is negative at certain hours. For time when correlation is negative, both uncertainties compensate each other's impact; hence overall impact would be less as compared to time when correlation is positive. Therefore, sale prices are adjusted in a way to exploit maximum robustness and opportuneness.

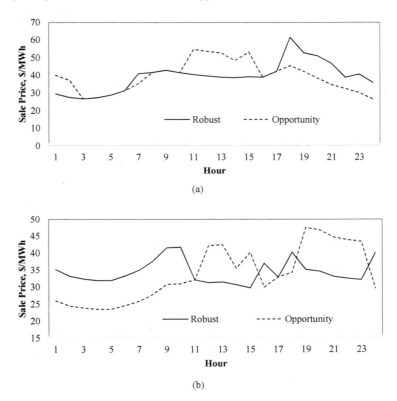

Fig. 6.9: Hourly sale price for (a) Weekdays and (b) Weekends

6.5.4.3 Impact of Demand Uncertainty

Fig. 6.10 indicates robustness and opportuneness curves considering (i) spot market price uncertainty alone and (ii) spot and demand uncertainties together. Dotted line represents robustness and opportuneness curves considering both uncertainties. Solid line represents these curves when only spot market price uncertainty is considered. As the dotted curve lies above solid line curve, this represents that for certain profit targets, tolerance of uncertainty is more when both the uncertainties are considered. Here consideration of demand uncertainty helps to increase robustness of the decision. However, in case of opportuneness, for certain anticipated profit, more uncertainty is required to achieve the same. This is because uncertain variations in demand and spot market prices compensate each other's impacts. Hence, demand uncertainty consideration is inevitable to achieve prudent decision making for LSE. It is worth to note here that results are reflective of specific data set and their nature would vary depending upon market conditions.

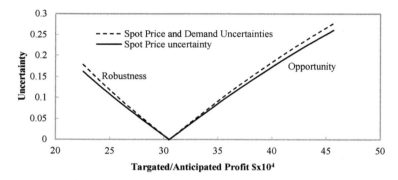

Fig. 6.10: Impact of demand uncertainty on robustness and opportuneness curves

6.6 Conclusions

The chapter proposes a novel model to maximize LSE's profit considering demand and spot market price uncertainties. A single framework is presented using IGDT approach. Ellipsoid-bound info-gap model is considered to model correlation of demand and spot market prices. Risk-averse and risk-seeking behavior of LSE is modeled by robustness and opportuneness functions. The proposed model is illustrated via case study of realistic Nord-Pool data, conducted for weekdays and weekends.

Results indicate that consideration of the two uncertainties in a single framework helps to improve robustness of a decision. LSE's decision become robust with lower profit targets as tolerance for uncertainty increases. Opportunity of high windfall profits is enabled when uncertain parameters are exposed to larger uncertainties. This study also highlights impact of demand uncertainty on robustness and opportuneness decision of LSE. It shows that tolerance of uncertainty is more when correlated uncertainties are considered.

Various models for LSE's decision making to maximize its profit are presented in Chapters 3-6. Next chapter of thesis presents the conclusions and future research directions.

REFERENCES

[1] J. Yang, J. Zhao, F. Luo, F. Wen, and Z. Y. Dong, "Decision-making for electricity retailers: A brief survey," *IEEE Trans. Smart Grid*, vol. 9, no. 5, pp. 4140-4153, Sept. 2018.

[2] S. A. Gabriel, M. Ferudun, and S. Balakrishnan, "A simulation approach to balancing annual risk and reward in retail electrical power markets," *IEEE Trans. Power Syst.*, vol. 17, no. 4, pp. 1050–1057, Nov. 2002.

[3] S. A. Gabriel, S. Kiet, and S. Balakrishnan, "A mixed integer stochastic optimization model for settlement risk in retail electric power markets," *Networks Spat. Econ.*, vol. 4, no. 4, pp. 323–345, Dec. 2004.

[4] C. K. Woo, R. I. Karimov, and I. Horowitz, "Managing electricity procurement cost and risk by a local distribution company," *Energy Policy*, vol. 32, no. 5, pp. 635–645, Mar. 2004.

[5] R. G. Karandikar, S. A. Khaparde, and S. V. Kulkarni, "Strategic evaluation of bilateral contract for electricity retailer in restructured power market," *Int. J. Electr. Power Energy Syst.*, vol. 32, no. 5, pp. 457–463, June 2010.

[6] A. J. Conejo, M. Carrión, and J. M. Morales, *"Decision Making Under Uncertainty in Electricity Market."* New York, NY, USA: Springer-Verlag, 2010.

[7] S. A. Gabriel, A. J. Conejo, M. A. Plazas, and S. Balakrishnan, "Optimal price and quantity determination for retail electric power contracts," *IEEE Trans. Power Syst.*, vol. 21, no. 1, pp. 180–187, Feb. 2006.

[8] J. M. Yusta, I. J. Ramírez-Rosado, J. A. Dominguez-Navarro, and J. M. Perez-Vidal, "Optimal electricity price calculation model for retailers in a deregulated market," *Int. J. Electr. Power Energy Syst.*, vol. 27, no. 5–6, pp. 437–447, July 2005.

[9] A. Hatami, H. Seifi, and M. K. Sheikh-El-Eslami, "A stochastic-based decision-making framework for an electricity retailer: Time-of-use pricing and electricity portfolio optimization," *IEEE Trans. Power Syst.*, vol. 26, no. 4, pp. 1808–1816, Nov. 2011.

[10] C. Defeuilley, "Retail competition in electricity markets," *Energy Policy*, vol. 37, no. 2, pp. 377–386, Feb. 2009.

[11] M. Mulder and B. Willems "Competition in retail electricity markets: An

assessment of ten years Dutch experience," June 2016. *Center Discussion Paper No. 2016-022.* [Online]. Available; https://ssrn.com/abstract=2796043.

[12] D. S. Kirschen, "Demand-side view of electricity markets," *IEEE Trans. Power Syst.,* vol. 18, no. 2, pp. 520–527, May 2003.

[13] R. Poudineh, "Liberalized retail electricity markets: What we have learned after two decades of experience?," *Oxford Institute for Energy Studies, United Kingdom,* Dec. 2019. [Online]. Availabe: https://ora.ox.ac.uk/objects/uuid:7f35 78a6-27b8-47d0-b2f2-c3aeff2b9 053

[14] S. Borenstein, "Time-varying retail electricity prices: theory and practice," in *Electricity Deregulation Choices Challenges,* J. M. Griffin and S. L. Puller, Ed., University of Chicago Press, Chicago, 2009, Ch. 8, pp. 315-378.

[15] A. Faruqui, R. Hledik, and J. Palmer, "Time-Varying and Dynamic Rate Design," *RAP Publications,* Bejing, July 2012. [Online]. Available: https://www.raponline.org.

[16] M. Zugno, J. M. Morales, P. Pinson, and H. Madsen, "A bilevel model for electricity retailers' participation in a demand response market environment," *Energy Econ.,* vol. 36, pp. 182–197, Mar. 2013.

[17] M. Liu, F. F. Wu, and N. Yixin, "A survey on risk management in electricity markets," *IEEE Power Eng. Soc. Gen. Meet.,* June 2006, pp. 1-6.

[18] M. Carrión, A. Conejo, and J. Arroyo, "Forward contracting and selling price determination for a retailer," *IEEE Trans. Power,* vol. 22, no. 4, pp. 2105-2114, Nov. 2007.

[19] G. Dutta and K. Mitra, "Dynamic pricing of electricity : A survey of related research," 2015. [Online]. Available: https://web.iima.ac.in/assets/snippets/ workingpaperpdf/1677509 22015-08-03.pdf.

[20] G. Dutta and K. Mitra, "A literature review on dynamic pricing of electricity," *J. Oper. Res. Soc.,* vol. 68, no. 10, pp. 1131–1145, Oct. 2017.

[21] A. Faruqui, R. Hledik, and J. Tsoukalis, "The power of dynamic pricing," *Electr. J.,* vol. 22, no. 3, pp. 42–56, Apr. 2009.

[22] P. Faria and Z. Vale, "Demand response in electrical energy supply: An optimal real time pricing approach," *Energy,* vol. 36, no. 8, pp. 5374–5384, Aug. 2011.

[23] A. Safdarian, M. Fotuhi-Firuzabad, and M. Lehtonen, "Impacts of time-varying electricity rates on forward contract scheduling of DisCos," *IEEE Trans. Power Deliv.,* vol. 29, no. 2, pp. 733–741, Apr. 2014.

[24] A. Safdarian, M. Fotuhi-Firuzabad, and M. Lehtonen, "A medium-term decision model for DisCos : Forward contracting and TOU pricing," *IEEE Trans. Power Syst.*, vol. 30, no. 3, pp. 1143–1154, May 2015.

[25] M. Khojasteh and S. Jadid, "Decision-making framework for supplying electricity from distributed generation-owning retailers to price-sensitive customers," *Util. Policy*, vol. 37, pp. 1–12, Dec. 2015.

[26] S. Nojavan, K. Zare, and B. Mohammadi-Ivatloo, "Risk-based framework for supplying electricity from renewable generation-owning retailers to price-sensitive customers using Information Gap Decision Theory," *Int. J. Electr. Power Energy Syst.*, vol. 93, pp. 156–170, Dec. 2017.

[27] R. Dahlgren and J. Lawarree, "Risk assessment in energy trading," *IEEE Trans. Power Syst.*, vol. 18, no. 2, pp. 503–511, May 2003.

[28] R. T. Rockafellar and S. Uryasev, "Conditional Value-at-Risk for general loss distributions," *J. Baking & Finance*, vol. 26, no. 7, pp. 1443–1471, July 2002.

[29] Y. Ben-Haim, *"Info-Gap Decision Theory, Decisions Under Severe Uncertainty."* California, CA, USA: Academic Press, 2006.

[30] M. Majidi, B. Mohammadi-ivatloo, and A. Soroudi, "Application of Information Gap Decision Theory in practical energy problems : A comprehensive review," *Appl. Energy*, vol. 249, pp. 157–165, Sept. 2019.

[31] J. Kettunen, A. Salo, and D. W. Bunn, "Optimization of electricity retailer's contract portfolio subject to risk preferences," *IEEE Trans. Power Syst.*, vol. 25, no. 1, pp. 117–128, Nov. 2009.

[32] N. Mahmoudi-Kohan, M. P. Moghaddam, M. K. Sheikh-El-Eslami, and E. Shayesteh, "A three-stage strategy for optimal price offering by a retailer based on clustering techniques," *Int. J. Electr. Power Energy Syst.*, vol. 32, no. 10, pp. 1135–1142, Dec. 2010.

[33] A. Ahmadi, M. Charwand, and J. Aghaei, "Risk-constrained optimal strategy for retailer forward contract portfolio," *Int. J. Electr. Power Energy Syst.*, vol. 53, no. 1, pp. 704–713, Dec. 2013.

[34] M. Shahidehpour, H. Yamin, and Z. Li, *"Market Operations in Electric Power Systems: Forecasting, Scheduling, and Risk Management."* Piscataway, NJ, USA: IEEE-Wiley-Interscience, 2002.

[35] G. Chicco, R. Napoli, P. Postolache, M. Scutariu, and C. Toder, "Customer characterization options for improving the tariff offer," *IEEE Trans. Power Syst*,

vol. 18, no. 1, pp. 381-387, Feb. 2003.

[36] M. Charwand, M. Gitizadeh, and P. Siano, "A new active portfolio risk management for an electricity retailer based on a drawdown risk preference," *Energy*, vol. 118, pp. 387–398, Jan. 2017.

[37] M. Khojasteh and S. Jadid, "Reliability-constraint energy acquisition strategy for electricity retailers," *Int. J. Electr. Power Energy Syst.*, vol. 101, pp. 223–233, Oct. 2018.

[38] S. Nojavan, K. Zare, and B. Mohammadi-Ivatloo, "Optimal stochastic energy management of retailer based on selling price determination under smart grid environment in the presence of demand response program," *Appl. Energy*, vol. 187, pp. 449–464, Feb. 2017.

[39] S. Nojavan, K. Zare, and B. Mohammadi-Ivatloo, "Selling price determination by electricity retailer in the smart grid under demand side management in the presence of the electrolyser and fuel cell as hydrogen storage system," *Int. J. Hydrogen Energy*, vol. 42, no. 5, pp. 3294–3308, Feb. 2017.

[40] S. Nojavan and K. Zare, "Optimal energy pricing for consumers by electricity retailer," *Int. J. Electr. Power Energy Syst.*, vol. 102, pp. 401–412, Nov. 2018.

[41] S. Nojavan, K. Zare, and B. Mohammadi-Ivatloo, "Application of fuel cell and electrolyzer as hydrogen energy storage system in energy management of electricity energy retailer in the presence of the renewable energy sources and plug-in electric vehicles," *Energy Convers. Manag.*, vol. 136, pp. 404–417, Mar. 2017.

[42] H. Golmohamadi and R. Keypour, "Stochastic optimization for retailers with distributed wind generation considering demand response," *J. Mod. Power Syst. Clean Energy*, vol. 6, no. 4, pp. 733-748, Jan. 2018.

[43] A. R. Hatami, H. Seifi, and M. K. Sheikh-El-Eslami, "Optimal selling price and energy procurement strategies for a retailer in an electricity market," *Int. J. Electr. Power Syst. Res.*, vol. 79, no. 1, pp. 246–254, Jan. 2009.

[44] J. Xu, P. B. Luh, F. B. White, E. Ni, and K. Kasiviswanathan, "Power portfolio optimization in deregulated electricity markets with risk management," *IEEE Trans. Power Syst.*, vol. 21, no. 4, pp. 1653–1662, Oct. 2006.

[45] S. J. Deng and S. S. Oren, "Electricity derivatives and risk management," *Energy*, vol. 31, no. 6–7, pp. 940–953, May 2006.

[46] S. Kharrati, M. Kazemi, and M. Ehsan, "Equilibria in the competitive retail

electricity market considering uncertainty and risk management," *Energy*, vol. 106, pp. 315–328, July 2016.

[47] D. S. Kirschen, G. Strbac, P. Cumperayot, and D. P. De Mendes, "Factoring the elasticity of demand in electricity prices," *IEEE Trans. Power Syst.*, vol. 15, no. 2, pp. 612–617, May 2000.

[48] M. J. Morey and L. D. Kirsch, "Retail Choice in Electricity : What Have We Learned in 20 Years ?," *Electric Markets Research Foundation*. 2016. [Online]. Available: https://hepg.hks.harvard.edu/files/hepg/files/retail_choice_in_electri city_for_emrf_final.pdf.

[49] R. De Sá Ferreira, L. A. Barroso, P. R. Lino, M. M. Carvalho, and P. Valenzuela, "Time-of-use tariff design under uncertainty in price-elasticities of electricity demand: A stochastic optimization approach," *IEEE Trans. Smart Grid*, vol. 4, no. 4, pp. 2285–2295, Apr. 2013.

[50] R. Garcia-Bertrand, "Sale prices setting tool for retailers," *IEEE Trans. Smart Grid*, vol. 4, no. 4, pp. 2028–2035, Dec. 2013.

[51] M. Aien, A. Hajebrahimi, and M. F. Fellow, "A comprehensive review on uncertainty modeling techniques in power system studies," *Renew. Sustain. Energy Rev.*, vol. 57, pp. 1077–1089, May 2016.

[52] M. Carrión, A. B. Philpott, A. J. Conejo, and J. M. Arroyo, "A stochastic programming approach to electric energy procurement for large consumers," *IEEE Trans. Power Syst.*, vol. 22, no. 2, pp. 744–754, Apr. 2007.

[53] A. Najafi-Ghalelou, S. Nojavan, and K. Zare, "Information Gap Decision Theory-based risk-constrained scheduling of smart home energy consumption in the presence of solar thermal storage system," *Solar Energy*, vol. 163, pp. 271–287, Mar. 2018.

[54] A. Akbari-dibavar, K. Zare, and S. Nojavan, "A hybrid stochastic-robust optimization approach for energy storage arbitrage in day-ahead and real-time markets," *Sustain. Cities Soc.*, vol. 49, pp. 1-9, Aug. 2019.

[55] S. Nojavan and K. Zare, "Interval optimization based performance of photovoltaic/wind/FC/electrolyzer/electric vehicles in energy price determination for customers by electricity retailer," *Sol. Energy*, vol. 171, pp. 580–592, Sept. 2018.

[56] M. Nazari and A. Akbari Foroud, "Optimal strategy planning for a retailer considering medium and short-term decisions," *Int. J. Electr. Power Energy*

Syst., vol. 45, no. 1, pp. 107–116, Feb. 2013.

[57] Z. Chen, L. Wu, and Y. Fu, "Real-time price-based demand response management for residential appliances via stochastic optimization and robust optimization," *IEEE Trans. Smart Grid*, vol. 24, no. 4, pp. 1-9, 1822-1831, Sep. 2012.

[58] A. Ehsan and Q. Yang, "State-of-art techniques for modelling of uncertainties in active distribution network planning: A review," *Appl. Energy*, vol. 239, pp. 1509–1523, Jan. 2019.

[59] A. Soroudi and T. Amraee, "Decision making under uncertainty in energy systems: State of the art," *Renew. Sustain. Energy Rev.*, vol. 28, pp. 376–384, Dec. 2013.

[60] J. R. Birge and F. Louveaux, "*Introduction to Stochastic Programming*," Chicago, Illinois, USA: Springer Science & Business Media, 2011.

[61] Y. Ben-haim, "Info-Gap Demand Theory : Economic Optimization Revisited Introduction Uncertainty in Economics," 2002. [Online]. Available: https://info-gap.technion.ac.il/files/2015/02/15econ_op.pdf

[62] Y. Ben-haim, "Uncertainty, probability and information-gaps," *Rel. Engg. and Syst. Safety*, vol. 85, no. 1-3, pp. 249–266, July 2004.

[63] A. Soroudi and M. Ehsan, "IGDT based robust decision making tool for DNOs in load procurement under severe uncertainty," *IEEE Trans. Smart Grid*, vol. 4, no. 2, pp. 886–895, June 2013.

[64] K. R. Hayes, S. C. Barry, G. R. Hosack, and G. W. Peters, "Severe uncertainty and Info-Gap Decision Theory," *Methods Ecol. Evol.*, vol. 4, no. 7, pp. 601–611, July 2013.

[65] M. Charwand and Z. Moshavash, "Midterm decision-making framework for an electricity retailer based on Information Gap Decision Theory," *Int. J. Electr. Power Energy Syst.*, vol. 63, pp. 185–195, Dec. 2014.

[66] S. Kharrati, M. Kazemi, and M. Ehsan, "Medium-term retailer's planning and participation strategy considering electricity market uncertainties," *Int. Trans. Electr. Ener. Syst.*, vol. 26, no. 5, pp. 920–933, May 2016.

[67] M. Khojasteh and S. Jadid, "A two-stage robust model to determine the optimal selling price for a distributed generation-owning retailer," *Int. Trans. Electr. Energy Syst.*, vol. 25, no. 12, pp. 3753–3771, Dec. 2015.

[68] S. Nojavan, H. Ghesmati, and K. Zare, "Robust optimal offering strategy of large

consumer using IGDT considering demand response programs," *Int. J. Electr. Power Syst. Res.*, vol. 130, pp. 46–58, Jan. 2016.

[69] A. Soroudi, A. Rabiee, and A. Keane, "Information Gap Decision Theory approach to deal with wind power uncertainty in unit commitment," *Int. J. Electr. Power Syst. Res.*, vol. 145, pp. 137–148, Apr. 2017.

[70] M. Mazidi, H. Monsef, and P. Siano, "Design of a risk-averse decision making tool for smart distribution network operators under severe uncertainties: An IGDT-inspired augment ε-constraint based multi-objective approach," *Energy*, vol. 116, pp. 214–235, Dec. 2016.

[71] J. Aghaei, M. Charwand, M. Gitizadeh, and A. Heidari, "Robust risk management of retail energy service providers in midterm electricity energy markets under unstructured uncertainty," *J. Energ. Engg.*, vol. 143, no. 5, pp. 1–14, Oct. 2017.

[72] P. Mathuria and R. Bhakar, "Info-gap approach to manage GenCo's trading portfolio with uncertain market returns," *IEEE Trans. Power Syst.*, vol. 29, no. 6, pp. 2916–2925, May 2014.

[73] K. Zare, M. P. Moghaddam, and M. K. Sheikh-El-Eslami, "Risk-based electricity procurement for large consumers," *IEEE Trans. Power Syst.*, vol. 26, no. 4, pp. 1826–1835, Nov. 2011.

[74] S. Nojavan, B. Mohammadi-Ivatloo, and K. Zare, "Robust optimization based price-taker retailer bidding strategy under pool market price uncertainty," *Int. J. Electr. Power Energy Syst.*, vol. 73, pp. 955–963, Dec. 2015.

[75] R. T. Rockafellar, "Optimization of Conditional Value-at-Risk," *J. Risk,* vol. 2, pp. 21–41, Sept. 2000.

[76] R. Schultz and S. Tiedermann, "Conditional Value-at-Risk in stochastic programs with mixed-integer recourse," *Math. Program.*, vol. 105, pp. 365–386, Feb. 2006.

[77] R. G. Karandikar, S. A. Khaparde, and S. S. Kulkarni, "Quantifying price risk of electricity retailer based on CAPM and RAROC methodology," *Int. J. Electr. Power Energy Syst.*, vol. 29, no. 10, pp. 803–809, Dec. 2007.

[78] T. Deng, W. Yan, S. Nojavan, and K. Jermsittiparsert, "Risk evaluation and retail electricity pricing using downside risk constraints method," *Energy*, vol. 192, pp.116672, Feb. 2020, doi: 10.1016/j.energy.2019.116672.

[79] M. Jamshidi, H. Kebriaei, and M. Sheikh-el-eslami, "An interval-based

stochastic dominance approach for decision making in forward contracts of electricity market," *Energy*, vol. 158, pp. 383–395, Sept. 2018.

[80] M. Denton, A. Palmer, R. Masiello, and P. Skantze, "Managing market risk in energy," *IEEE Trans. Power Syst.*, vol. 18, no. 2, pp. 494–502, May 2003.

[81] W. Steinhurst, D. White, A. Roschelle, A. Napoleon, R. Hornby, and B. Biewald, *Portfolio Management: Tools and Practices for Regulators.* 2006. [Online]. Available; http://www.synapse-energy.com/

[82] Y. Liu and X. Guan, "Purchase allocation and demand bidding in electric power markets," *IEEE Trans. Power Syst.*, vol. 18, no. 1, pp. 106–112, Feb. 2003.

[83] J. M. Yusta, H. M. Khodr, and A. J. Urdaneta, "Optimal pricing of default customers in electrical distribution systems: Effect behavior performance of demand response models," *Int. J. Electr. Power Syst. Res.*, vol. 77, no. 5–6, pp. 548–558, Apr. 2007.

[84] M. Carrión, J. M. Arroyo, and A. J. Conejo, "A bilevel stochastic programming approach for retailer futures market trading," *IEEE Trans. Power Syst.*, vol. 24, no. 3, pp. 1446–1456, May 2009.

[85] C. Zhao, S. Zhang, X. Wang, X. Li and L. Wu, "Game analysis of electricity retail market considering customers' switching behaviors and retailers' contract trading," *IEEE Access*, vol. 6, pp. 75099–75109, Nov. 2018.

[86] N. Mahmoudi-Kohan, M. P. Moghaddam, and M. K. Sheikh-El-Eslami, "An annual framework for clustering-based pricing for an electricity retailer," *Int. J. Electr. Power Syst. Res.*, vol. 80, no. 9, pp. 1042–1048, Sept. 2010.

[87] M. G. Lijesen, "The real-time price elasticity of electricity," *Ener. Econo.*, vol. 29, pp. 249–258, Mar. 2007.

[88] A. Knaut and S. Paulus, "Hourly price elasticity pattern of electricity demand in the German day-ahead market," No. 16/07. EWI Working Paper, 2016. [Online]. Available; https://www.econstor.eu/handle/10419/144865.

[89] A. Knaut and S. Paulus, "Real-time price elasticity of electricity demand in wholesale markets," *Sustainable Energy Policy and Strategies for Europe, 14th IAEE European Conf.*, Rome, Italy, Oct. 2014. [Online]. Available; https://www.iaee.org/proceedings/article/12406.

[90] X. Labandeira, J. M. Labeaga, and X. López-otero, "A meta-analysis on the price elasticity of energy demand," *Energy Policy*, vol. 102, pp. 549–568, Mar. 2017.

[91] B. Neenan and J. Eom, "Price elasticity of demand for electricity: A prime and

synthesis," *Electr. Power Res. Inst.*, Jan. 2008. [Online]. Available; https://www.epri.com.

[92] S. Sekizaki and I. Nishizaki, "Decision making of electricity retailer with multiple channels of purchase based on fractile criterion with rational responses of consumers," *Int. J. Electr. Power Energy Syst.*, vol. 105, pp. 877–893, Feb. 2019.

[93] A. J. Conejo, J. M. Morales, and L. Baringo, "Real-time demand response model," *IEEE Trans. Smart Grid*, vol. 1, no. 3, pp. 236–242, Oct. 2010.

[94] B. Colson, P. Marcotte, and G. Savard, "Bilevel programming : A survey," *4or*, vol. 107, pp. 87–107, June 2005.

[95] B. Colson, P. Marcotte, and G. Savard, "An overview of bilevel optimization," *Ann. Oper. Res.*, vol. 153, no. 1, pp. 235–256, Sept. 2007.

[96] E. Alekseeva and L. Brotcorne, "A bilevel approach to optimize electricity prices," *Yugoslav J. Oper. Res.*, vol. 29, no. 1, pp. 9-30, Feb. 2018.

[97] S. Sekizaki, I. Nishizaki, and T. Hayashida, "Electricity retail market model with flexible price settings and elastic price-based demand responses by consumers in distribution network," *Int. J. Electr. Power Energy Syst.*, vol. 81, pp. 371–386, Oct. 2016.

[98] R. Shari, S. H. Fathi, and V. Vahidinasab, "A review on demand-side tools in electricity market," *Renew. and Sust. Energ. Rev.*, vol. 72, pp. 565–572, May 2017.

[99] N. O. Connell, P. Pinson, H. Madsen, and M. O. Malley, "Benefits and challenges of electrical demand response : A critical review," *Renew. and Sust. Energ. Rev.*, vol. 39, pp. 686–699, Nov. 2014.

[100] V. S. K. Murthy Balijepalli, V. Pradha, and S. A. Khaparde, "Review of demand response under smart grid paradigm," *ISGT2011-India*, Kollam, Kerala, 2011, pp. 236-243.

[101] N. G. Paterakis, O. Erdinç, and J. P. S. Catalão, "An overview of demand response: Key-elements and international experience," *Renew. Sust. Energ. Rev.*, vol. 69, pp. 871–891, Mar. 2017.

[102] A. Meyabadi Fattahi and M. H. Deihimi, "A review of demand-side management: Reconsidering theoretical framework," *Renew. Sust. Energ. Rev.*, vol. 80, pp. 367–379, Dec. 2017.

[103] D. Yu *et al.*, "Modeling and prioritizing dynamic demand response programs in

the electricity markets," *Sust. Cities Soc.*, vol. 53, pp. 1-9, Feb. 2020.

[104] F. C. Schweppe, M. Caramanis, R. Tabors, and R. Bohan, "*Spot Pricing of Electricity.*" Chicago, Illinois, USA: Springer Science & Business Media, 2013.

[105] S. Nojavan, R. Nourollahi, H. Pashaei-didani, and K. Zare, "Uncertainty-based electricity procurement by retailer using robust optimization approach in the presence of demand response exchange," *Int. J. Electr. Power Energy Syst.*, vol. 105, pp. 237–248, Feb. 2019.

[106] N. Mahmoudi, M. Eghbal, and T. K. Saha, "Employing demand response in energy procurement plans of electricity retailers," *Int. J. Electr. Power Energy Syst.*, vol. 63, pp. 455–460, Dec. 2014.

[107] J. C. do Prado and W. Qiao, "A stochastic decision-making model for an electricity retailer with intermittent renewable energy and short-term demand response," *IEEE Trans. Smart Grid*, vol. 10, no. 3, pp. 2581-2592, Feb. 2018.

[108] A. Asadinejad and K. Tomsovic, "Optimal use of incentive and price based demand response to reduce costs and price volatility," *Int. J. Electr. Power Syst. Res.*, vol. 144, pp. 215–223, Mar. 2017.

[109] M. Nasouri and G. Alfred, "A two-stage stochastic framework for an electricity retailer considering demand response and uncertainties using a hybrid clustering technique," *Iran. J. Sci. Technol. Trans. Electr. Eng.*, vol. 9, pp. 541-558, July 2019.

[110] G. Gutiérrez-Alcaraz, J. H. Tovar-Hernández, and C. N. Lu, "Effects of demand response programs on distribution system operation," *Int. J. Electr. Power Energy Syst.*, vol. 74, pp. 230–237, Jan. 2016.

[111] Y. Yang, "Understanding household switching behavior in retail electricity market," *Energy Policy*, vol. 69, pp. 406–414, June 2014.

[112] E. Çelebi and J. D. Fuller, "A model for efficient consumer pricing schemes in electricity markets," *IEEE Trans. Power Syst.*, vol. 22, no. 1, pp. 60–67, Jan. 2007.

[113] E. Celebi and J. D. Fuller, "Time-of-Use pricing in electricity markets under different market structures," *IEEE Trans. Power Syst.,* vol. 27, no. 3, pp. 1170–1181, Feb. 2012.

[114] M. Doostizadeh and H. Ghasemi, "A day-ahead electricity pricing model based on smart metering and demand-side management," *Energy*, vol. 46, pp. 221–230, Oct. 2012.

[115] F. H. Moghimi and T. Barforoushi, "A short-term decision-making model for a price-maker distribution company in wholesale and retail electricity markets considering demand response and real-time pricing," *Int. J. Electr. Power Energy Syst.*, vol. 117, pp. 1-18, May 2020.

[116] L. Zhao, Z. Yang, and W. J. Lee, "The impact of time of use (TOU) rate structure on consumption patterns of the residential customers," *IEEE Trans. Indus. Appl.*, vol. 53, no. 6, pp. 5130-5138, Nov.-Dec. 2017.

[117] J. Mays and D. Klabjan, "Optimization of time-varying electricity rates," *The Energ. J.*, vol. 38, no. 5, pp. 1–38, June 2017.

[118] C. Dong, C. T. Ng, and T. C. E. Cheng, "Electricity time-of-use tariff with stochastic demand," *Prod. Oper. Manag.*, vol. 26, no.1, pp. 64-79, Jan. 2017.

[119] J. Yang, J. Zhao, and F. Wen, "A framework of customizing electricity retail prices," *IEEE Trans. Power Syst.*, vol. 33, no. 3, pp. 2415-2428, Sept. 2017.

[120] C. Feng, Y. Wang, K. Zheng, and Q. Chen, "Smart meter data-driven customizing price design for retailers," *IEEE Trans. Smart Grid*, Oct. 2019. doi: 10.1109/TSG.2019.2946341.

[121] J. P. Chaves-Avila, K. Wurzburg, T. Gomez, and P. Linares "The green impact: How renewable sources are changing EU electricity prices," *IEEE Power and Energ. Mag.*, vol. 13, no. 4, pp. 29–40, June 2015.

[122] C. Zhu, R. Fan, and J. Lin, "The impact of renewable portfolio standard on retail electricity market: A system dynamics model of trpartite evolutionary game," *Energy Policy*, vol. 136, pp. 1-13, Jan. 2020.

[123] H. Yi, M. H. Hajiesmaili, Y. Zhang, M. Chen, and X. Lin, "Impact of the uncertainty of distributed renewable generation on deregulated electricity supply chain," *IEEE Trans. Smart Grid*, vol. 9, no. 6, pp. 6183–6193, May 2017.

[124] E. Kyritsis, J. Andersson, and A. Serletis, "Electricity prices, large-scale renewable integration, and policy implications," *Energy Policy*, vol. 101, May 2016, pp. 550–560, Feb. 2017.

[125] E. Trujillo-baute and P. Mir-artigues, "Analysing the impact of renewable energy regulation on retail electricity prices," *Energy Policy*, vol. 114, pp. 153–164, Mar. 2018.

[126] Competetion and Consumer Protection Commision *"Review of Competition in the Electricity and Gas Retail Markets,"*, 2017. [Online]. Available; https://www.ccpc.ie/business/wp-content/uploads/sites/3/2017/05/Submission-

References

to-the-CER-Review-of-Competition-in-the-Electricity-and-Gas-Retail-Markets. pdf.

[127] M. Zarif, M. H. Javidi, and M. S. Ghazizadeh, "Self-scheduling approach for large consumers in competitive electricity markets based on a probabilistic fuzzy system," *IET Gen. Trans. Distr.*, vol. 6, no. 1, pp. 50-58, Jan. 2012.

[128] "PJM Website." 2020. [Online]. Available: https://www.pjm.com/markets-and-operation/energy.aspx.

[129] S. Pfenninger and I. Staffell, "Long-term patterns of European PV output using 30 years of validated hourly reanalysis and satellite data," *Energy*, vol. 114, pp. 1251–1265, Nov. 2016.

[130] "The GAMS Website." 2020. [Online]. Available: https://www.gams.com/.

[131] "GAMS Solver Information." 2020 [Online]. Available: https://www.gams.com/latest/docs/S_MAIN.html.

[132] M. Zugno, J. M. Morales, P. Pinson, and H. Madsen, "Modeling demand response in electricity retail markets as a stackelberg game," *12th IAEE European Energy Conference: Energy challenge and environmental sustainability*, Venice, Italy, 2012, pp. 1-11.

[133] B. Drysdale, J. Wu, and N. Jenkins, "Flexible demand in the GB domestic electricity sector in 2030," *Appl. Energy*, vol. 139, pp. 281–290, Feb. 2015.

[134] A. Mohammadi and S. Rabinia, "A comprehensive study of Game Theory applications for smart grids, demand side management programs, and transportation networks," in *Smart Microgrids*, S. Bahrami and A. Mohammadi, Ed., Cham, Switzerland: Springer, 2019, pp. 57-64.

[135] A. Sinha, P. Malo, and K. Deb, "A review on bilevel optimization : From classical to evolutionary approaches and applications," *IEEE Trans. Evol. Comp.*, vol. 22, no. 2, pp. 276-295, June 2017.

[136] L. Bird, M. Milligan, and D. Lew, "Integrating variable renewable energy : Challenges and solutions," National Renew. Energy Lab, Golden, Co. USA, Tech. Rep. NREL/TP-6A20-60451, Sept. 2013.

[137] J. Fortuny-Amat and B. McCarl, "A representation and economic interpretation of a two-level programming problem," *The J. Oper. Res. Soc.*, vol . 32 , no . 9, pp. 783–792, Sept. 1981.

[138] H. Rastegar, S. H. Hosseinian, B. Vahidi, and H. Khazaei, "Two-level decision-making model for a distribution company in day-ahead market," *IET Gen. Trans.*

Distr., vol. 9, no. 12, pp. 1308–1315, June 2015.

[139] "Solar Data." 2020. [Online]. Available: https://www.renewables.ninja/.

[140] A. Mehdizadeh, N. Taghizadegan, and J. Salehi, "Risk-based energy management of renewable-based microgrid using Information Gap Decision Theory in the presence of peak load management," *Appl. Energy*, vol. 211, pp. 617–630, Feb. 2018.

[141] "Nord Pool Website," 2020. [Online]. Available; https://www.nordpoolgroup.com/historical-market-data/.

[142] A. Soroudi, *"Power System Optimization Modeling in GAMS."* Switzerland: Springer International Publishing, 2017.

APPENDIX-A

Operational Cost Linearization of Thermal Generating Unit

According to Fig. A.1, linearization of operational cost of thermal generation is formulated as follows [143]

$$0 \leq P_{g,t}^k \leq \Delta P_g^k u_{g,t} \quad \forall k = 1:n \tag{A.1}$$

$$\Delta P_g^k = \frac{P_g^{\max} - P_g^{\min}}{n} \tag{A.2}$$

$$P_{g,ini}^k = (k-1)\Delta P_g^k + P_g^{\min} \tag{A.3}$$

$$P_{g,\text{fin}}^k = \Delta P_g^k + P_{g,ini}^k \tag{A.4}$$

$$P_{g,t} = P_g^{\min} u_{g,t} + \sum_k P_{g,t}^k \tag{A.5}$$

$$C_{g,ini}^k = a_g \left(P_{g,ini}^k\right)^2 + b_g P_{g,ini}^k + c_g \tag{A.6}$$

$$C_{g,\text{fin}}^k = a_g \left(P_{g,\text{fin}}^k\right)^2 + b_g P_{g,\text{fin}}^k + c_g \tag{A.7}$$

$$s_g^k = \frac{C_{g,\text{fin}}^k - C_{g,ini}^k}{\Delta P_g^k} \tag{A.8}$$

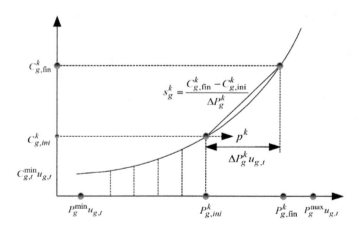

Fig. A.1: Cost curve of generating units [142].

Printed in the USA
CPSIA information can be obtained
at www.ICGtesting.com
LVHW020827210923
758625LV00014B/848